John Cleland

Evolution, Expression and Sensation

Cell life and pathology

John Cleland

Evolution, Expression and Sensation
Cell life and pathology

ISBN/EAN: 9783337095482

Printed in Europe, USA, Canada, Australia, Japan

Cover: Foto ©berggeist007 / pixelio.de

More available books at **www.hansebooks.com**

EVOLUTION, EXPRESSION,
AND SENSATION.

PUBLISHED BY

JAMES MACLEHOSE, GLASGOW.

—

MACMILLAN AND CO., LONDON.

London, ... *Hamilton, Adams and Co.*
Cambridge, . . . *Macmillan and Co.*
Edinburgh, . . . *Douglas and Foulis.*

—

MDCCCLXXXI.

EVOLUTION, EXPRESSION,

AND SENSATION,

CELL LIFE AND PATHOLOGY.

BY

JOHN CLELAND, M.D., F.R.S.,

PROFESSOR OF ANATOMY IN THE UNIVERSITY OF GLASGOW.

GLASGOW:
JAMES MACLEHOSE, ST. VINCENT STREET,
Publisher to the University.
1881.

CONTENTS.

INTRODUCTION.

THIS volume is not intended exclusively for either medical men or biologists, but for all who take an interest in the modern speculations inseparably bound up with the present position of biological science.

The first five of the six articles now published together, although they have been written at different periods and with different objects in view, are devoted to subjects more or less cognate. So that one will be found to illustrate allusions made in others.

The conception which it is sought to defend in the address "On the Evolutions of Organization" is that these evolutions are definite, and that the highest evolution of animal life is completed in man. Development both in the individual and in the totality of life is not only a development from a simple beginning, but a development towards a completed whole. There is morphological design, and when in any line of development the design is completed, the evolution ceases, although, by the operation of environment or external circumstances, variations may continue to occur and degenerations of diverse kinds may take place.

Such views demand for the universe a background or underlying element of spirit. Among the evidences of the place occupied by spirit in nature, I count that which is afforded by what I term symbolic correlation as highly important, and the demonstration of the existence of symbolic correlation I have sought to establish by an analysis of human expression. At the conclusion of the article on that subject the existence of such a principle of expression as conducing to the characters of biological evolution is simply hinted, while it is referred to more fully, but still with great brevity, in the first article of the series.

Under the head of "Vision" it is shewn that the evolution of sense-organs is a very different thing from the evolution of sensation; and while it is pointed out that natural selection furnishes no adequate explanation of the rise of the organ of vision as a structure, attention is called to the circumstance that the very existence of vision and the other senses points to there being an unknowable territory whence, and not from the material world, they take their origin.

In discussing the subject of vision, the whole subject of sensation has naturally come up, and reference is made to the doctrine of sensation put forward in the memoir " On the Physical Relations of Consciousness and the Seat of Sensation." That doctrine may not have received much attention ; but it is in the position of remaining unassailed, while the old doctrine is unsatisfactory to some of the most competent judges, and remains undefended from the objections here brought against it.

With regard to the fifth article of the series, it need only be said that the connection of the theory of cell-life with that of life in every other aspect is too close to require more than mention here ; while to one who believes in life within life and in the unity of cause in the order of events in each, cell-life is especially interesting as the simplest of a series of which the most complex known is the life-evolution on the face of the earth.

Every one is familiar by this time with the re-luctance of certain physicists and naturalists to take into consideration even the possibility of such an element as spirit being necessary for the construc-tion of a rational philosophy of nature. The wonderful advances in physics, affording sure foot-ing for further progress within that domain, and giving play for speculations formerly inconceivable, may lead some physicists to overleap those strict fences the observance of which has secured the ad-vance of their favourite studies, and they may dream that the physical is the only world. So also

there are biologists so startled by the progress
made through physical methods in explaining
phenomena once deemed inexplicable, save by reference to an agency interfering with physical laws,
that they think they have a glorious revelation
before them when they account life to be nothing
more than a chance result of the operation of those
laws to which the manifestations of physical energy
are no doubt as subject within the body as in the
inorganic world. Such an attitude is natural in
the enthusiasm of discovering that so much is explained which was formerly set aside as inexplicable
because vital. The argument runs much in this
fashion : Vitality not only gave no explanation of
any of the operations within the body, but its
alleged virtues prevented the search for explanation;
but physical research has explained much, and we
may expect it to explain much more. So far well.
But they add that it must explain everything ; and
in this they go too far. I trust to show that this
notion, often put forward with much dogmatism and
with unnecessary rancour, must be set aside because
there are phenomena, such as morphological plan,
which cannot possibly be referred at any future time
to physical laws, but indicate spirit.

To those working habitually with physical
methods, no doubt spirit seems a vague and intangible entity because unsubject to those methods.
The very word spirit brings down to us from olden

times the record of the difficulty which is still experienced in grasping the non-material, since its original meaning, breath, is nothing more than movement of the least tangible form of matter. But this difficulty of looking out on spirit as an object, without imagination stepping in to invest it dimly with material form, has its counterpart in the impossibility of looking at material phenomena as if from within, without attributing to them a consciousness like our own : and this also has had no trifling effect on the thought of primitive nations. The disadvantage of our position in relation to matter is similar to that which we suffer in respect of spirit ; matter we know only from without, spirit only from within.

But, further, our consciousness is plainly a mere surface-wave on an inscrutable deep ; and brought to that surface there is perpetual evidence of effort, and thus the known properties of spirit are consciousness and force. On the other hand, the phenomena of matter are confined to movement in space, one particular form of force ; it presents but a finite exhibition of one of the known properties of spirit, while it shows no sign of the other, namely, of consciousness. Therefore, matter gives us no explanation of spirit ; but in the hidden ocean of infinite spirit is locked the secret of matter.

While, however, matter in its relations to matter presents merely one order of the properties of spirit,

posited in points or centres, and each centre acting
on the others, when we review the nervous system
in its relations with consciousness we pass from the
mere mutual relations of such centres, and find both
mind acting on matter, and matter acting on mind.
Those who deny the evidence of spirit existing independent of matter point to the permanence of energy
in the material world.   They argue that the mind
is affected by physical stimuli, and that no mental
operation takes place without corresponding physical action.   They assume that, were consciousness
beyond the circle of physical changes, it would re-
main unaffected by them, instead of mind acting on
body, and body on mind, as is the case ; but if it be
within the circle of physical change then, they argue,
it is itself physical, or at least we have no right to
suppose that it can exist independent of matter.
From the dogma that mind cannot exist independent of matter, one can easily see how simple is the
passage to some such formula as that mind is an
evolution of matter.   But to extract a clear meaning from such a formula is not so easy : it is the
endowing of matter with properties demonstrated in
no physical nor chemical laboratory.   Thus when
in the passage quoted at p. 23, Häckel speaks of
spirit as a function of force he uses a form of words
which conveys no meaning.   Had he said that force
is a function of spirit he might have reached a truly
" monistic" philosophy and thrown the mind back on

infinite spirit as the source of all finitudes, instead of compelling it to rest on an unfounded imagination that the particles of matter had other than their known properties and that these were sufficient for the evolution of consciousness.

As to the exact position of the mind in relation to the chain of physical changes in the body, we may in the future learn more about it than we know at present, but, meanwhile, it is quite certain that mental action is not in quantitative relation to the preceding external influences applied to the organs of sense. If a servant whispers in your ear that there are robbers in the house, there will be caused much less vibration of the drum of your ear and consequent action of the auditory nerve than by the loud ringing of a dinner bell ; but there will possibly result very much greater mental disturbance. The stimuli in both cases would be applied to the same nerves ; and no physical theory can represent it as possible that the channels taken in the brain by the irritation conveyed along the nerves would vary according to the meaning of the sounds. It is plain therefore that the physical stimulant in sensation does not lie in the same relation to the mental changes immediately following, as does the charge exploded in a gun to the flight of the bullet. The vibration in the ear produces change in the nerves proportionate to its amount : and science does not know how the last of these changes is related to consciousness so as to affect it. The next change, so variable in amount, due neither to vibrations nor to sense of sound but to the idea which the sound symbolizes, begins in the mind ; and science does not know how it happens that it is accompanied with change in the brain, of strictly proportionate amount.

These are undeniable facts, though not what a confiding public has been always taught in science lectures. It may be added that there is no experimental evidence to show whether the amount of nervous change resulting from a given amount of physical stimulus is ever the same after passing through a nervous centre in which it affects consciousness as it would be after passing a centre in which it does not. But we have every reason to believe that the reflex actions in which no consciousness is involved are strictly proportional to the amount of stimulus ; while we know that when consciousness is affected it will often happen that no obvious action will follow a stimulus which would otherwise be sufficient to produce movements.

In defending the position that the evolutions observable in organization are definite, I have used in the address which follows, few and simple illustrations, but I trust they are sufficient to give stability to the argument. Manifestly the more complex problems of morphology could not have been made use of, as they would have necessitated too many technicalities, and disputes on innumerable points of detail. Take for example, the structure of the vertebrate skull. Nothing, to my thinking, could illustrate morphological design more beautifully. But painstaking workers have found the subject exceedingly difficult, and the majority of anatomists have at different times been contented to accept the dicta of some authority on it, instead of studying nature for themselves. The opinions disseminated in that fashion among English students during the last sixteen years, originating in rather perfunctory observations, involve in my opinion the grossest misinterpretations; and unless these are cleared away the morphological beauty of the skull cannot be seen. I content, myself, therefore, with taking this opportunity of referring students who may wish to know my views on this subject to a memoir "On the relations of the Vomer, Ethmoid and Inter-maxillary bones," published in the Philosophical Transactions, 1862; and to an address in the British Association Reports, 1875; and await a more favourable occasion to discuss its intricate details.

The graduation address, which is the last of the articles brought together in this volume, claims no close relationship with the others; yet the remarks on truth which it contains may possibly be of service to those engaged in the abstract questions to which these articles are devoted, as well as to the young practitioners for whom they were written. I do not suppose for a moment that they are less required by the upholders of one view than by adherents of any other. In republishing that address I desire especially to direct attention to the unsatisfactory state of the law with regard to everything affecting our knowledge of the causes of death.

I.

THE EVOLUTIONS OF ORGANIZATION.

A

I.

THE EVOLUTIONS OF ORGANIZATION.

(An Address delivered at the Opening of the Medical Classes in the University of Glasgow, 26th October, 1880.)

THE study of medicine presents to him who would view it properly two great aspects :—primarily, it exhibits an art for the relief of the sufferings of others ; but it also displays a wide field of investigation for the satisfaction and development of the inquirer's own mind.   So evidently is this the case, and so naturally does the study of the body tend to a wider survey of the realms of nature, that in ancient, as well as more modern times, research of the most extensive character, and far-reaching speculations, have occupied the attention of physicians.   As soon as the student begins his curriculum, by seeking a knowledge of the human body he comes into the presence of the great system of organization on the face of the earth, of which the structure of man forms the crowning object, not to be justly appreciated save in conjunction with the rest.   As he becomes acquainted

with the phenomena of life, and learns the mechanical and chemical processes involved in them,
he naturally asks how far the laws of life are to
be explained by the laws of matter, and what is
the motive power of the order of phenomena both
in the individual and in the series.   There come up
for his consideration the links uniting to the body
that spiritual element which, in the actions of the
individual, is never seen by us in operation, save
in the closest association with physical structure.
And the yet larger speculation, Is spirit or matter
the prior, the underlying, unchangeable, and eternal
element, constantly obtrudes itself on the horizon
of more limited inquiries.

These are all questions which have been much
discussed in recent times ; and the order of the
appearance of structures having been studied both
in the form of palæontology and development, it is
not wonderful that speculations have been rife as
regards not only the nature, but likewise the origin
of life, of species, of matter, and intelligence.

That there is a certain unity binding together
the most diverse forms of organization is a doctrine
which, at this date, I fancy no one will deny.   The
similarity of the units of life in plant and animal
texture, and the general, if not altogether universal,
pervasion of sexual distinction, are illustrations of

that; and if I confine my view to the animal kingdom, I think the days are past when any one will consider it chimerical to compare vertebrates with invertebrates, or the most dissimilar invertebrates one with another. At once it will be granted that ova, however various, are comparable, and that it is matter for observation to compare the stages of their growth as they give rise to forms that are far asunder. Nor will any serious doubt be entertained on survey of that highest and most important assemblage of animals the vertebrata, that they have appeared on earth in the order of their complexity; however great may be the mystery in which the rise of invertebrata may be wrapped. All this involves the conception of a complex unity acquiring its complexity stage by stage, even as the individual develops from the ovum to the adult condition; and such a conception may be justly termed Evolution.

But such an evolution may be conceived of variously, both in respect of character and cause. In its character it may be conceived of as a growth without aim, forming altogether an indefinite aggregation like the sum of the branches of a tree; or the view may be held that it is an orderly arrangement, like some vast temple in which every minaret and most fantastic ornament has got its own appointed place and harmonies, while the central tower ascends to its pre-ordained completion,—an evolution like that of a plant from whose root-leaves the shoot ascends with leaves that change their character according to law, till the summit is reached, where definite groups converge, assorted to complete the flower.

And just as there are diverse notions as to the character of the evolutions of life on the globe, so there is difference of opinion as to the source from which it springs. There are some who still deny the genetic connection of different forms of life, and others who either consider genetic connection proved, or look on it as a hypothesis with more or less probability in its favour ; and among these latter there are some who see no morphological facts which cannot be explained by reference to genetic relationship and its involved lines of heredity; while there are others like myself who, in heredity, can only recognize a phenomenon the origin of which demands an explanation.

It is a very remarkable circumstance that while all these views are held and have been known to the scientific world for a great length of time, the name of *Evolutionists* has, with curious obliviousness, been assumed as a distinctive title by those who believe that the evolution is merely indefinite and entirely to be explained by heredity. It is for them to say if this exactness in the choice of a name is an index of the accuracy of the reasoning on which their views are founded. Had they called themselves Demolitionists, on account of their disbelief in morphological design, the name might possibly have been more expressive.

Looking, however, at evolution in the natural sense of the word, we may find it useful to give a glance to some of the more notable opinions that have been held on the subject.

In such a survey, however brief, the school of Schelling, as represented by Oken and Carus, must not be overlooked. Professor Häckel of Jena claims for Oken a foremost place in the pedigree of that system which he has himself put forward ; and, undoubtedly, though there is little other affinity between Häckelism and the doctrine which Oken published at the same University of Jena in his "Naturphilosophie" in 1810, than that both draw largely on the imagination, Oken's is a theory of evolution embracing all Nature. It is built upon an *a priori* conception. It constantly declares that things must be after a certain fashion, simply because the conception demands it ; and it is scarcely surprising that, proceeding by this method, not distinguishing between speculation and demonstration, its details are often far astray. His own words are :—" I am very well aware that there is many an object which does not stand in its right place, but where again is there a single system in which this is not more strikingly the case ? We have here dealt only with the restoration of the edifice, wherein, after years of long and oft-repeated attempts, the furniture may for the first time be properly distributed, without detriment to its general bearings or ground plan." [1]

That which made Oken's writings a living power was, that to him nature was in all its parts replete with meaning, inspired with an inherent fitness, and not the mere chance result of a concourse of atoms unaccounted for.   Laying the foundation of his system in a mathematical conception, he seeks to ally himself not to Epicurus but to Pythagoras. "Every real," he says, "is absolutely nothing else than a number. This must be the sense entertained of numbers in the Pythagorean doctrine, namely, that everything or the whole universe had arisen from numbers. . . . The essence in numbers is naught else than the Eternal. The Eternal only is or exists, and nothing else is when a number exists." [2] He lays claim to having advanced, so

[1] Elements of Physiophilosophy, translated for the Ray Society by Alfred Tulk, p. xiv.   [2] Op. cit. p. 13.

early as 1805, the doctrine that all organic beings
originate in and consist of cells, which constitute the
protoplasm from which all larger organisms are
evolved;[1] and he is more generally recognized as
having been the first to perceive that the skull consists of segments in serial continuity with those of
the vertebral column. But principally it concerns
us to notice that while Oken appreciated the correspondence between the ovum, the beginning of
life in the complex animal, and the "oozoa" or
simplest forms of animals, he saw in the animal
world an unity completed in man; or, to use
his own words, the animal kingdom is only a
dismemberment of the highest animal, *i.e.*, of
man."[2]

The plan sketched by Oken was, so far as the
animal kingdom was concerned, modified and
elaborated by C. G. Carus. Guided, like Oken, by
the aphorism that the whole is repeated in every
part, Carus published with his beautiful work on
Comparative Anatomy a volume of "Researches
on Philosophic or Transcendental Anatomy," as
perplexing a puzzle as could well be conceived for
a plain observer of Nature. But it is to be noted
among the much that is good which we owe to him,
that he recognized as a cardinal fact the central

[1] Op. cit. p. xi.   [2] Op. cit. p. 494.

position of the digestive system throughout the animal kingdom, the truth at the bottom of Häckel's gastræa theory; and that he so far appreciated the relation of the segmented invertebrata to the vertebrata as to perceive that, while in certain lower forms the nervous system completed a circle round the mouth, in the invertebrate segmentata the nervous centres were concentrated on the under, and in the vertebrata on the upper side; seeing in this what Oken had already observed through the organic world, and what so skilled and accurate an observer as Dana[1] has more recently recognized in this very matter, a system of expression in nature, according to which things of highest dignity are placed uppermost.

In France, Etienne Geoffroy St. Hilaire, claiming to be free from all *a priori* fancies, but acknowledging the influence which the definition of Leibnitz had on him, that "the order of the universe is variety in unity," was led to the conception of the "unity of organization" by the results of researches into points of detail; and collecting those detailed researches in one work, his "Philosophie Anatomique," he commences by asking, "Can the organization of vertebrate animals be reduced to one

[1] J. D. Dana "On Cephalization," &c.   Ann. and Mag. Nat. Hist., Sept., 1863.

type?"[1] and subsequently arrives at the doctrine that "all animals are made on one and the same type." Founding on observation, this great anatomist appreciated not only the correspondence of part with part in different animals, but the existence of inherent laws of symmetry and order governing the formation of structures similar in different animals and hereditarily transmitted. To express it in the distinct nomenclature of Owen, he recognized the existence of general as well as special homologies. The compeer of Lamarck, he saw reason, even before that writer, to judge that the dogma of immutability of species, in all time and circumstances, lacked proof ; and, at a period later than Lamarck's work, he cast it aside, thus affording evidence to those who may require it that the appreciation of the orderly internal laws of organization is not inconsistent with a full appreciation of the possibility of a genetic connection of diverse forms — seeing that, in this instance, one independent thinker originated for himself both ideas[i] and clearly perceived that the second could in no degree allow the first to be dispensed with. His notion was, that the "ambient world having undergone changes from one geological epoch to another, even the atmosphere having varied in chemical

---

[1] Philosophie Anatomique, II., p. 445,

composition, and the conditions of respiration having been thus modified, actual forms had to differ in organization from their ancestors of ancient times, and that according to the degree of the modifying force." [1]

Lamarck, who next claims attention, has given the summary of his doctrine in language so brief and clear, that I may quote his own words. He lays down four principles :—" 1. Life, by its own forces, tends continually to increase the volume of the whole body which possesses it, and to extend the dimensions of its parts up to a limit which it determines. 2. The production of a new organ in an animal body results from a new want supervening, which continues to make itself felt, and from a new movement which this want gives birth to, and continues. 3. The development of the organs and their force of action are constantly in ratio of the employment of these organs. 4. All that has been acquired, laid down or changed, in the organization of the individuals during the course of their life, is conserved by their generation and transmitted to the new individuals proceeding from those which have undergone those changes." [2]

[1] Mémoires de l'Academie Royale des Sciences, XII., 63.   Isidore Geoffroy St. Hilaire, Histoire Naturelle Générale, II., p. 416.

[2] Lamarck, Hist. Nat. des Animaux sans Vertèbres. 3rd Edition, p. 57.

The proofs of the effect of new wants or appetency, in producing new organs, he derives from
the acknowledged effect of habitual action in increasing the development of organs. But while it is
difficult to limit the results of use, disuse, or peculiar
action in developing, dwindling, or modifying structures already in some form existing, it is not easy
to understand how such a principle could operate
to produce an organ the basis of which was not
previously laid down.

That which gives the air of absurdity to Lamarck's
illustrations is that, altogether unlike St. Hilaire, he
appears to have ignored the necessity for a principle independent of external circumstances to regulate form, in, for example, such matters as symmetry.
The following may be given as an instance : — " I
conceive that a gasteropod mollusc, which, as it
crawls along, finds the need of touching the bodies
in front of it, makes efforts to touch those bodies
with some of the foremost parts of its head, and
sends to these every time quantities of nervous
fluids, as well as other liquids. I conceive, I say,
that it must result from this reiterated afflux towards
the points in question that the nerves which abut
at these points will, by slow degrees, be extended.
Now, as in the same circumstances other fluids of
the animal flow also to the same places, and espe-

cially nourishing fluids, it must follow that two or more tentacles will appear and develop insensibly in those circumstances on the points referred to. This is, without doubt, what has happened in all the races of gasteropods whose wants have given rise to the habit of touching bodies with the parts of their head."[1] Thus he allowed himself to pass in the most guileless way, like many a subsequent writer, from "*je conçois*" to "*sans doute.*"

Lamarck espoused the doctrine of spontaneous generation, as he was bound logically from his point of view to do. His contention was that it was improbable that under the government of the material universe by secondary causes there should be a deviation from that system either in the first appearance or subsequent evolution of life ; and he failed, erroneously, as I believe, to see that in the phenomena of life any element was present different in kind from the phenomena of dead matter. "Certes," he says, "the power which has made the animals has made them all that they are, and endowed them with the faculties observed in each, by giving an organization fitted to produce them. Observation authorizes us to recognize that this power is *Nature*, and that she is the product of the will of the Supreme Being, who has made her what

[1] Op. cit. p. 59.

she is."[1]   Observe how unjustifiable are the charges which have been brought against him from a theological point of view. "Chose étrange!" he exclaims afterwards, "l'on a confondu la montre avec l'horloger, l'ouvrage avec son auteur."[2]

In 1844 there was published in Edinburgh the book entitled "Vestiges of the Natural History of Creation," a work which appears to have been treated with scant justice by any party. Darwinians have thought it necessary to disparage both this work and Lamarck; while to those defenders of Christianity who have so little faith in its own strength that they think it cannot stand without their coopering, and smell danger perpetually from afar, it mattered nothing that the author of the "Vestiges," as well as Lamarck, emphatically declared the necessity, according to his view, for a Creator. He described himself modestly as " a private person with limited opportunities of study"; and in the wide area over which he ranges he sometimes seeks support from things which an expert in this or that department would touch with a wary hand ; but independence and originality, temperance and frankness are qualities which he displays in an unusual degree.

The reasons which led him to think that spon-

[1] Op. cit. p. 66.   [2] Op. cit. p. 95.

taneous generation had, at least at a former period
of the earth's history, existed, were the same as
influenced Lamarck; but his hypothesis of the
mode in which variety and complexity have been
reached is different, namely, " that the several
series of animated beings, from the simplest and
oldest, up to the highest and most recent, are,
under the providence of God, the results, *first*, of
an impulse which has been imparted to the forms
of life, advancing them in definite times, by generation through grades of organization terminating
in the highest dicotyledons and vertebrata, these
grades being few in number, and generally marked
by intervals of organic character, which we find to
be a practical difficulty in ascertaining affinities;
second, of another impulse connected with the vital
forces, tending in the course of generations to
modify organic structures in accordance with external circumstances, as food, the nature of the
habitat and the meteoric agencies, these being the
'adaptations' of the natural theologian."[1]   Thus,
while he saw the possibility of a modifying power
being exerted by external circumstances, he appreciated clearly the necessity of recognizing an
element on which it was exerted; and the necessity
for that "impulse," as he calls it, ought to have

[1] Vestiges of the Nat. Hist. of Creation, 10th edition, p. 155.

warned him from adhesion to spontaneous generation. But I hold him to be right in considering
the progress from simpler to more complex life in
the history of the earth as being analogous to the
development of an individual from the embryonic
to the adult condition, a definite, and not an indefinite evolution.

Such were the various views involving an evolution in one sense or another with which biology
was familiar before the year 1859, when Darwin's
Origin of Species made its appearance. In that
work there is full agreement with Lamarck in respect that the origin of life from a creator is frankly
and continually referred to, and that the question
brought up is essentially the old one of the degree
in which forms through long epochs of time are
capable of changing; but spontaneous generation
is let alone, and the author is content to express
his belief that "animals have descended from at
most only four or five progenitors, and plants from
an equal or lesser number."[1] The new element
imported into the discussion is, as you are aware,
what he terms "natural selection," a result of the
struggle for existence. This struggle for existence
had already been pointed out by Owen[2] as a potent

[1] Origin of Species, 3rd edition, p. 518.
[2] Zoological Transactions, vol iv, p. 15. See, also, Comparative Anatomy of Vertebrates, vol. iii, p. 799.

factor in extinction of species, but it was left for
Darwin to originate the idea that it could do what
Owen recognized as due to innate causes, viz., that
it could produce variety and advance by the continual survival in the contest of the "fittest," *i. e.*
of those which by accidental variations of structure
had in the struggle an advantage over their neighbours.   The struggle exists, beyond all question ;
and few can read Darwin's works without being
impressed with the likelihood of its resulting during
long periods in an indefinite amount of variety.
Yet, even as to this, it may be remarked that there
is a tacit assumption in the theory, that individuals
surviving by slight variations of different kinds
would be separated one from the other, so as not
to mix the different peculiarities to which they
severally owed their survival; and it may be objected that such separation would be unlikely so
long as the variations were yet exceedingly slight ;
while it may be judged probable from Mr. Darwin's
own researches that individuals with slight consan-
guinity and differing minutely one from the other,
but not too greatly, would be exactly those which
would be most prolific and so give rise to a mixing
of incipiently diverging characters.

The more important criticism can in no way be
eluded that the doctrine of natural selection has

precisely the same defect as the Lamarckian doctrine of appetency, only to a greater degree; it
does not account for the formation of any new
organ, nor for new organs appearing symmetrically.
No doubt, on the "je conçois" principle it may be
made to lengthen and strengthen to any extent
any number of structures already existing, and,
potent for degeneration also, may be supposed to
dwarf others, when they become incumbrances.
Thus it is noticeable that in the series of forms
preceding the limbs of the horse, a story on which
so much is sometimes founded, no new structure
makes its appearance; simply the third digits have
enlarged in size, while those on each side have
become smaller; and in the horse of the present
day both the enlargement and the dwindling have
reached a degree beyond which it is difficult to
conceive them passing. I notice these circumstances though I am not prepared to dogmatize to
the effect that it actually was by the sole agency
of natural selection that the series of limb-forms
alluded to found an appropriate completion in the
horse. But that natural selection should give rise
to totally new and symmetrical organs is hard to
imagine and impossible to prove. Mr. Darwin, to
a certain extent, acknowledged the difficulty, and
boldly he launched an attempt to account for the

organ of vision, on the principle that if so complex a structure could be disposed of, no difficulty need be made about any other. Starting from a nerve-extremity coated with pigment as the simplest in "the great kingdom of the Articulata" he sees no difficulty "in believing that natural selection has converted the simple apparatus of an optic nerve merely coated with pigment and invested by transparent membrane into an optical instrument, as perfect as is possessed by any member of the great Articulate class."[1]   Then, without attempting to explain the vertebrate eye, which, in the individual, is developed like many other structures, through a series of forms utterly incapable of function, he tells the reader that, if he finds the other facts in the book bear out the theory, he is bound to admit that the eye of an eagle might be formed by natural selection. No doubt, the evolution of the vertebrate from the invertebrate eye is made easier by Kowalevsky's discoveries in the larval ascidians ; but what is called an eye in amphioxus is only a spot without mechanism for vision, so that it is really true that the invertebrate eye is blotted out before the true vertebrate eye, which, in the individual, passes through so many visually useless stages of development, suddenly appears before

[1] Op. cit. pp. 206 and 207.

us, explain it as you will, in full functional operation. But even supposing, as seems possible
enough, that vertebrates possessing vision were
once linked to forms like larval ascidians by animals of soft structure with gradually increasing
perfection of sight, and that amphioxus is after
all degenerated from some member of such a lost
chain of transitional forms, the visually useless
stages of development of the vertebrate eye
cannot be the mere accumulated record of a
series of gradually improving optical instru-
ments.

No more than Lamarck has Darwin considered
that it is not a sensitive nerve alone which is
required to begin vision or any other special sense,
but a capability of the consciousness to be modified
in a way altogether incomparable with the equally
incomprehensible affection which constitutes general sensation.

While the Darwinian system adds the idea of
natural selection to the stock of hypotheses for
explanation of evolution by external influences, it
denies the existence of any definite evolution of
organization dependent on a definite cause. While
it has the greatest faith in the power of the ovum
to carry down the most minutely determined
details of future development, it denies that there

is anything apart from the accidents of external
circumstances to direct the paths of the whole
world-history of life, and thus shuts out of consideration the whole class of phenomena which
were not only built on by Oken, but were patent
to St. Hilaire, and Owen, and Goodsir; and it
not only leaves both sex and symmetry unaccounted
for, but renders them inexplicable, although they
are matters obvious to every one, and pervading the
whole organic world.

It is worthy of notice that the position taken,
with admirable moderation, by Darwin himself,
that living forms have originated from a very few
progenitors, is no aid to the conception of continuity, unless there be added the doctrine of
spontaneous generation,—the absence of evidence
for which has been the subject of an honest and
pathetic wail from an eminent Darwinian. Either
the assumed primordial forms sent onward, in the
stream of heredity to future ova, a something not
derived from external circumstances, or they owed
all their properties to the operation of the laws of
dead matter. If they were the accidental evolutions of dead matter,—that is spontaneous generation. If they owed their properties to another
source, then every ovum that exists has similar
properties distinct from those of dead matter, as

believers in definite evolution or morphological design contend.

Professor Häckel, of Jena, founding on the Darwinian hypothesis, not only pushes it to the furthest limit, but accounts for every thing by what he calls the "monistic philosophy."

According to the materialists, says he, the matter has originated the force; and, according to the spiritualists, the force has originated the matter; but both are dualistic and wrong: according to the monistic philosophy you can neither think of force without matter, nor of matter without force. "Spirit and soul are only higher and combined or differentiated powers of the same function, which we indicate by the most general expression as force; and force is an universal function of all matter."[1] He has interested himself and his readers inventing most detailed pedigrees of different forms, more or less valuable according to the degree in which they express established morphological affinities, and recommends, to the consternation of so able a scientific and political authority as Virchow, that this sort of thing should be taught in the public schools of Germany. He talks with admiration of Oken and St. Hilaire as his predecessors in evolution, and fails to see that

[1] Anthropogenie, pp. 707-8.

they were on another line of rails, and moving in an opposite direction. Affinity means morphological relationship, and is a matter of observation ; but the pedigree of species cannot, in the nature of things, be the subject of either observation or induction. Every affinity pointed out is a scientific gain ; but it is no advantage to science to put affinities forward in the form of an allegation that the allied forms are descended from a common ancestor of imaginary structure. If evolutions are definite, or, in other words, if morphological design exist, the necessity for explaining *all* affinity by genetic relationship disappears.

The quotation from Häckel just made will probably lead you to reflect that it is of the utmost importance, in considering the relations of both life and intelligence to matter, to have a clear understanding what the properties of matter are. Now, if we consider the phenomena of matter in its inorganic forms, we cannot fail to see that they are all of one order,—viz., related to movement in space. The energy spoken of in connection with matter is nothing but movement and tendency to movement,—a thing whose dealings are altogether with locality. But it is the immediate experience of every one that consciousness is not movement in space. No doubt it is found indirectly by science that consciousness does not exist in organisms without a proportionate amount of molecular action ; but that is no reason for confounding the one thing with the other, and saying that consciousness *is* molecular action,—a statement which conveys no meaning.

And just as consciousness is not to be confused with the molecular actions associated with it in organisms, so life is something else than the sum of the chemical and mechanical operations engaged in its manifestation. Its phenomena are no doubt mere movement in space, and therefore allied to those of matter rather than to those of consciousness, but they exhibit two closely connected characters—development and heredity, to which we find nothing analogous in the inorganic world. Crystallization is not analogous to development, for crystallization is uniform in its intimate parts ; whereas the essence of development is the sequence of a definite series of forms entirely differing one from another, and heredity is the transmission of this remarkable property. While, then, the experiments of Pasteur and Lister conclusively prove that there is no such thing as spontaneous generation in the present; the difference in character between development and the laws of matter makes spontaneous generation an inadequate hypothesis to account for the first appearance of life in the past.

Much less shall we feel inclined to such a notion, if we discern in the whole train of life on the face of the earth a series of developments in different directions, as determined as the development of the ovum into an adult individual.

But, it will be said, what of the law of continuity? It is for the sake of continuity that spontaneous generation has been sought after. If, it is argued, the evolutions of the movements, forms, and conditions of the heavenly bodies have happened by the mere laws of matter, is it likely that there has been the intervention of miracle in peopling this world of life? Well, as I read the series of evolutions, we must begin where Plato began, with the Eternal and Unlimited. Matter, the lowest grade of the evolving cosmos in which we are placed, is an exhibition of that divine energy acting from points in only one manner, namely, movement in space; and those centres of movement only exist as such one for another. In that case, matter is not eternal, but is itself an intrusion into the field which it occupies, followed in the history of this globe by other intrusions in which all the continuity existing is that they occur gradually. We do not know how gradually matter came to its present condition; but

if the whole area of space within our cognizance
gives no hint of a development from which sprang
the separate chemical elements so-called, what likelihood is there that the span of their changes in
time could come within our recognition? The
phenomena peculiar to life, and summed up in
development, we know have appeared by slow
degrees; and intelligence, an addition totally different in kind, both from matter and from life,
makes its appearance likewise by insensible gra-
dation.

If the development of each individual be related
to a larger development as the lives of the tissue-
elements are related to the life of an organism, and
if that larger development proceeds in definite directions to termini or adult conditions, I own that I
can see nothing out of harmony. It has seemed to
some of the ablest biologists that have ever lived that
there is a vast amount of evidence that such termini exist. But the prejudiced superstition that
starts with the dogma that there can be no design
nor aught that is complete in nature, because the
laws of matter must be capable of accounting for
everything, never deigns to consider the evidence of
ordered evolution or completeness, but sets it aside
with a sneer as "scholastic nonsense" and "arche-
typal follies," and calls that attitude the "method

of modern science." Rather, I apprehend, the method of true science, modern or not, would be to study each evolution separately, attending to all the phenomena, and investigating its limits, and the action of every possible factor in its production. Thus, truly, it is of importance to study the operation of circumstances of environment at the present day, that we may understand the effects of environment in the past; and science is enormously indebted to Darwin for the stimulus which he has given to that inquiry. But just as in the evolution of language, although environment plays an all-important part, the fundamental factors are the existence of ideas to communicate and a parallelism of things of different order, rendering it possible for ideas to be represented by signs, so also in the evolutions of organization there is a non-material element in the definitely directed impulse, and there is a physiognomic propriety in organic forms, a class of facts appreciated by both Oken and Dana, telling of spirit which pervades the whole.

It is certain that the evolutions of form after passing through periods of activity do cease; for while all the most curious forms of invertebrate life,—polyzoa, echinodermata, lamellibranchs, brachiopoda and cephalopoda, are formed already in palæozoic times, the vertebrates have in the same circumstances of geological change been most active since then, and passed from piscine forms on to man. And neither the structure nor the intellect of man surpasses now the perfection that it had reached in ancient Egypt and in Greece ; though the lapse of time has proved sufficient for variations and degenerations. The definite lines of development on which the head had gradually risen to the perfection exhibited by the classic sculptors are incapable of being carried further : the face is curved in under the skull so far that it could not be carried back to a greater extent, and leave room for teeth, tongue, and throat.

Nor can attention be too frequently directed to the ordered and completed evolution seen in the history of the heart, a remarkable series from the simple to the complex, but showing in the amphibian and reptilian stage a more complex mechanism, yet less perfect machine than in the fishes, so that, as I have elsewhere said, "it might have been difficult to explain if it could have been noted by an observer before birds and mammals appeared on the earth."[1] But the mechanism gradually developed is completed, with variety, in the bird and mammal, and shows no sign of undergoing further complication.

[1] Animal Physiology, p. 116.

One often hears final causes spoken of with a contempt which is indeed only a revulsion from a style of writing which will not now find many admirers, in which adaptations were found by pointing out what extraordinary consequences would follow some impossible alteration in nature, and final were made to do the duty of efficient causes; but in the history of the vertebrate heart may be seen a remarkable instance of the definite evolution of a complex mechanism to perform a particular kind of work. There is no reason to doubt that here we have morphological evolution, and final cause combined; just as it is possible to imagine, though we may have little experience of it, a building morphologically belonging to the Gothic order, yet teleologically fitted for the wants of modern science.

It is a legitimate position to take up, that all the evolutions of nature are definite, but that the series of such evolutions is indefinite in number and kind; that individual evolutions, like other individuals, are finite, but form members of a larger total. So, in the evolutions of organization we see vortic units, the textural elements, receiving and rejecting currents of material, while they maintain during a finite life-time their individuality, and these united into larger individuals subject to the same law. In the evolution of vegetation, which everywhere exhibits definite individuals, or organs, structurally united in indefinite series, the highest groups are composed of compound individuals in whose structure the most complex, and in every way highest evolution occurs in connection with the highest vegetable function, the perpetuity of the series of finite individuals; and, the flower once perfected, evolution has ceased, though variation continues. In the animal kingdom, as has been long appreciated, even from the ovum the part devoted to vegetal function has its surface turned inwards, and the complexity of the animal sphere, which is wrapped around the vegetal, takes place in connection with the conscious faculties. By the ministry of sensation and voluntary movement the food is introduced into the specially vegetal parts, while in connection with the perfecting of those animal functions a process takes place such as in the vegetable kingdom is seen only in the flower, namely, that the portions, whether segments or organs, become more definite in number, more fused and compact. Thus, the largest feature in the evolution of both plants and animals consists in that which is simple passing into the manifold, and the manifold being compacted together in a higher unity. These evolutions are subjected in the vegetable kingdom to the service of nutrition and reproduction; while in the animal kingdom nutrition and reproduction give way in importance to the development of intelligence. And, however imperfectly zoologists may yet agree as to the evolutions in detail in different parts of the animal kingdom, it is plain that in the human form an organism has at last appeared constituting an abode of intelligence such as exists in no other, and that in man alone intelligence reaches the capability of ascending beyond the wants of the physical organism in the contemplation of abstract truth.

This organism, I repeat, has not improved with the progress of discovery in modern science, but was at least as complete in the heroes of antiquity as in those of recent times. It is surely, then, an assumption to suppose that evolution as distinguished from variation in animal forms must go on unchecked till astronomic change shall have ended the capability of this world to support life. It is far more probable that the evolutions of the future are to be sought in realms with which the zoologist acknowledges that he has nothing to do, and take origin out of the special psychical characters of man.

For my own part I maintain that the universe,

the "πρὸς τὸν Θεὸν," presents an endless unrolling of definite evolutions, and that the evolutions of organization are completed, though its variations continue, while other evolutions are at work, and more will appear.

II.

ON THE ELEMENT OF SYMBOLIC CORRELATION IN EXPRESSION.

II.

ON THE ELEMENT OF SYMBOLIC CORRELATION IN EXPRESSION.[1]

THE very use of the word expression implies a relationship between mind and body; for that which is expressed is a condition of mind, and that by which it is expressed is a condition of body; while the problem remains for both the naturalist and the metaphysician—By what means do movements of the body, or more widely, conditions of matter afford an index to conditions of the mind?

Expression may be said to be conveyed through the medium of the senses of sight and hearing. The other senses may be left out of consideration; for flavours and odours, however far-reaching their effects on the percipient, have no utility whatever in directly determining conditions of other minds, and the sense of touch refers to forms and movements

[1] Originally published in the Journal of Anatomy and Physiology, July, 1879.

better determined usually by sight. By the blind, forms and movements are appreciated through touch, which by others are more quickly perceived through the medium of vision ; and in the case of the deaf, visible signs may be made to serve a purpose better fulfilled by words when words can be heard ; but it remains true that expression is a mechanism of forms, appreciable movements, and sounds, and that these are most generally conveyed through the portals of eye and ear.

Thus the problem of expression as I have defined it involves the whole study of the origin of language, and the same gulf has to be bridged over in determining how meanings have become attached to words as in determining how they are attached to arrangements of feature and gesture. But the origin of the primitive symbols in speech is so obscure, and the interaction of circumstances so complex in the elevation of them into languages, and in determining the differences and changes of these, that one can hardly expect as yet any further light to be thrown by philology on expression by feature and gesture than that which is afforded by the mere recognition of the fact that language is a symbolic mechanism, to be grouped along with everything else to which the term expression can be applied.

Such a recognition, however, does afford the reflection that since such a complex system of symbols has certainly arisen within the limits of man's existence, and has not been inherited from any ancestry among the lower animals, it is unnecessary to suppose that the far simpler language of feature and gesture has been so inherited, even in those instances in which similar movements occur in man and other animals.

It is also instructive that the action of individuals in initiating language is infinitesimal, that the art of speech is acquired by observation and imitation in which the learner is largely unconscious of the details of the process which he imitates, and that he learns the meaning of words simply by observing their constant association with ideas otherwise expressed, but not from any appreciation of inherent connection between the words and the ideas; any such connection having been in most instances completely disguised long ages ago in the changes through which the words have passed. This being palpably the case, it is not surprising that into the simpler language of feature and gesture an imitative element largely enters, nor wonderful that it presents much whose origin is difficult to account for.

Permanent Expression.

The most recent, and probably the most profound and elaborate attempt to unravel the mysteries of expression is Mr. Darwin's, and I think it is to be regretted that so acute and original an observer has confined himself so strictly to the expression of the emotions, and neither allowed his mind to diverge to the expression of thought by language, nor to that permanent expression due to form of skeleton and chiselling of soft parts. Of course it is open to any one to take up the position that there is no relationship between the characteristics of the mind and the permanent forms of the body, and such a position is often maintained. But even if it be granted, as it well may be, that conclusions derived from bodily conformation are often liable to be delusive, and even if the extreme view be held that the forms of the features and other parts of the body never afford any key whatever to mental qualities, it yet remains incontrovertible that the mind of the observer is so affected by different bodily forms as to associate them with different mental qualities, and to feel a sense of the unexpected when convinced in any instance that the association is violated by nature. Thus it is one of the acknowledged aims of the artist to convey by forms permanent characters of mind.

It may, indeed, be fairly held that a graceful form, by awakening the idea of the graceful in the onlooker, will lead him to associate that idea with the possessor of the form, and will tinge his judgment of the mind behind it; especially when the graceful forms are found in the head, the seat of the organ of mental action, and the face the special index of its changes. So also the idea of strength given by a well-knit body may be erroneously allowed to impress the judgment favourably as to the presence of strength of mind ; and, indeed, it may also be held that the habits engendered by a sturdy or weakly bodily development have much influence in modifying mental tone. But no such considerations will account for those various symbolisms of form by means of which every competent painter is able to portray minds of various mould, apart from the temporary expression. I do not attempt at the present moment to explain why ; but assuredly very different mental characteristics will be indicated by varying the proportion of breadth to length in drawing a face, or by varying the proportion of one-third of the face to the rest. A massive chin is so distinctly a physiognomic representation of firmness, that an artist would in vain attempt to exhibit the resolution of a Cromwell in a face with a small and narrow jaw, or with one of those pretty chins like a bagatelle ball, not uncommon in certain localities. And the chin is the more remarkable as a feature of expression since its projection forwards is distinctive of the higher races of humanity, and has nothing to do with the muscles of mastication in connection with which the lower jaws of savages are often heavy and strong. It has no physical function whatever, so far as I am aware. None of the lower animals have a vestige of it, and the lower races of humanity approximate to them in this respect, without loss of power either in the lips or jaw. Curiously it happens that the *levatores menti*, muscles correctly named by the old anatomists *musculi superbi*, adapted by pushing the lower lip upwards to aid that closure of the mouth by which is expressed a resolve to resist, accompanied with an assertion of superiority,—it happens that those muscles render the integuments of the chin more prominent; and if they were sufficiently strong, and if the integuments to which their fibres descend were adherent to the bone, their habitual action would increase the forward prominence; but both they and the other muscles of the face attached to the chin are far too small for the wildest fancy to suppose that they can possess such an influence; therefore their action, instead of affording an explanation of the chin, rather exhibits a parallel problem in expression.

Another very striking illustration of the suggestion of mental character by permanent bodily form is to be obtained thus :—Taking a profile sketch, leave the features unaltered, but make additions or subtractions from the occipital region and back of the neck by changing the line descending from the position where the occipital and parietals meet, or from a lower point. Grave subtractions from the part so bounded become incompatible with the expression of mental stability long before they are carried to such an extent as to be anatomically improbable; but it must not be forgotten that the change so simply sketched implies the gravest alteration in the form of the brain and skull in their whole extent. Here, then, we have an instance of change of expression produced by the altered form of parts whose primary function is certainly not one of expression, and thus contrasting in a marked way with the changes producible by form of chin.

In the present day, however, the theory of Gall and Spurzheim is justly exploded; and I make bold to repeat what I have already stated elsewhere, that the vague and helpless notions of localization of mental functions in different parts of the cerebral hemispheres, so fashionable with so-called "medical psychologists," have no support from the facts of comparative anatomy, pathology, and experiment, all of which show that the hemispheres have a function common to the mass of their grey matter, so far as thought is concerned. Therefore, it would appear that differences of shape of head, not involving differences of cranial capacity, when they indicate differences of mental character, have a purely physiognomic value perfectly similar to that of differences of features of the face. We may even go further, and admit that it is probable that many of the statements of the followers of Gall have a large amount of physiognomic truth although their theory is utterly wrong.

These remarks are, I think, sufficient to illustrate that in the expressiveness of permanent forms of the body a class of phenomena exists, not to be explained by reference to the "three principles" which appear to Mr. Darwin "to account for most of the expressions and gestures involuntarily used by man and the lower animals under the influence of various emotions and sensations," viz. : "Serviceable associated habits," "the principle of antithesis," and "the principle of actions due to the constitution of the nervous system, independently of the will." [1]

I might have alluded more particularly to those forms which depend not on the skeleton but on the soft parts, giving the presence or absence of chiselling to the features ; the term chiselling indicating curves such as suggest a firm material moulded into shape, as contrasted with those into which soft pulp might gravitate. But it may well be answered that such modelling being pleasant to the eye is inherited by artificial selection, and that, though on this account more common in the educated classes, it may be largely present in the absence of grace or culture of mind or heart, while in other instances these latter may be present, and modelling of the features absent. A fair argument might be sustained that the circumstances favourable to moral and mental selection are often coincident with those favourable to physical selection ; and that, apart from this, it is natural to associate the pleasant in mind with the pleasant in body, forms noble on account of mere physical harmonies with nobility of moral description, even though the two things may not be associated in the external world. Yet there are probably few who will doubt that if two children of the same parents, closely resembling

[1] Darwin, Expression of the Emotions, p. 28.

one another, be taken, the one left in neglect, the other subjected to educating influences, the difference of treatment will be likely to tell on the moulding of the features.

Expression of the Emotions.

The element in expression of emotion, whether by gesture or feature, which appears to me to be the most important, has often been entirely overlooked ; and although it seems to have been largely present to the mind of Piderit and of Gratiolet,[1] yet the rationale has not been effectively expounded, and we see one of the most elaborate and ingenious writers on the subject, Mr. Darwin, throwing it altogether aside.

I shall now try to put it in definite form. To this end, I observe first, that words indicating position and quantity represent ideas relating to both the physical and mental world. Secondly, emotions expressible by such words are indicated by the attitudes, gestures, and movements of body expressed by the same words. Thirdly, the same principle is applicable to movement of feature.

1. For the sake of clearness, it may be well to discuss the first of these propositions separately. The words useful for our purpose by referring to

[1] Darwin, loc. cit., p. 6.

both physical and mental conditions are such as the following :—Those expressing height, as—upward and downward, ascent and descent, elevation and depression, superiority and inferiority, rise and decline or fall, over and under; those expressing other directions, as—forward and backward, advance and retrogression, before and behind or to one side, direct and roundabout, straight and oblique ; those expressing distance, as—far and near, approach and separation, attraction and repulsion ; words expressing magnitude, as—large and small, wide and narrow, expanded and contracted : words expressing resistance, as—strong and weak, hard and soft, firm and yielding ; words connected with motion or rest, as—quick and slow, tension and relaxation.

The connection between the physical and other meanings of these words is in most instances not far to seek. No doubt at first sight it may seem puzzling to find anything in common between moral elevation and physical elevation or mere distance from the earth's surface, and one may look on it as remarkable that in the ideas of all men the two things are associated, and are so by a link independent of the peculiarities of individual languages, so that one is led to suspect that the bond is not only universal but necessary;

and a similar difficulty may be felt at the application of such words as "advance," not to movement in space but to conditions of other than physical description.

But very little consideration will make such difficulties disappear. They do not exist in relation to many of the words enumerated above; for magnitude, resistance, and motion, three of the heads used in the enumeration, are obviously ideas not confined to objects in space, though primarily the words refer to physical objects, according to the general rule of words with a physical and metaphysical application. As regards words expressing distance, it will be noted that "near" or "far" remains the same idea, whether the degree of deviation from identity contemplated refers to time, space, or constitution; while attraction or repulsion is the mere tendency to passage from one degree of nearness to another, whether in space or in constitution.

In such expressions as "advance" and "retrogression" we have still to deal with distance; only in this instance it is distance from a goal more or less distinctly imagined as an object of desire, whether it be a spot on earth to which we bend our steps, or some intellectual, moral or artistic perfection, or whatever else; while the character of the means adopted to reach the goal is expressed by such words as " direct " or " roundabout."

There remains only the metaphoric use of words referring to height to be explained of all the list with which we began. Here another principle comes into play : we have not to do with a common idea of various application, but with two distinct ideas associated by circumstance. Because the sky and the natural sources of light are above, because vegetable life attracts attention most by growth above the surface of the ground, because the visible products of putrefaction sink down and become buried over, because dead bodies fall, because in activity we stand up, and in rest lie down, and because a lofty position commands attention and gives physical advantages, therefore a host of associations grow up in the human mind, by which " upward " represents the good, the great, and the living, " downward " the evil and the dead. In the same manner, indeed more directly, we associate impressions through the organs of sense with impressions from the moral world similarly pleasant or otherwise, as in the case of sweetness, bitterness, brightness and gloom.

2. Emotions to which such words as we have been considering have application are indicated by the attitudes, gestures, and movements of body

expressed by the same words. More shortly, the workings of the mind are expressed by attitudes, gestures, and movements of body of a nature correlative with them.

That which we like we desire to be near to, what we dislike we seek to avoid ; but it is not merely on these accounts that we bend the body forwards and approach that which pleases us, while we retreat or draw our head and body back from what is offensive. In numerous instances such movements and gestures are made not from any notion of achieving a purpose, and still less from an inherited habit founded in their utility to real or supposed ancestors, but simply from the close connection subsisting between movement towards an object and mental attraction to it, or between movement away from an object and a feeling of repulsion.

A similar remark holds good with regard to movement of the arms, which perform gestures of receiving and rejecting. It may be mentioned in passing that so far as these are performed from the shoulder, they are accomplished respectively by the *pectoralis major* muscle, which might be termed the muscle of embrace, and the *latissimus dorsi* muscle, which might be called the muscle of rejection. Lift the arm into the position which places the

latissimus dorsi on the greatest stretch, and sweep
it downwards and backwards, with the palm turned
away from the body, and no gesture of the limb
can more thoroughly express the putting away of
something vile. Nay, more, if the same movement
be carried out also by the forearm and hand, the
gesture begins with the palm in front of the face as
if to conceal from the eyes what is loathsome, and
passes from this to the removal of it altogether.
Yet it is a gesture applied to the intangible and
invisible ; by it the cleric puts away false doctrine,
and the fastidious sublimely brands a notion as
vulgar.

In like manner, slight movements of the arms
express the hugging of an idea to the bosom
when nothing but what is thoroughly impersonal is
thought of, and the fingers bend as if to keep a
something in the hand when nothing but delightful
sentiment is concerned. Thus, one may frequently
see among children at play, when an amusement is
proposed, the right or more active arm thrown upwards and inwards towards the opposite shoulder,
and the hand gently closed, while the word " come "
is on the lips, and that when no removal to another
place is intended. It is partly the expression of a
wish for all to join in concert, partly it expresses
the pleasure with which the object to be joined in is

anticipated. Similarly, if an artist wished to express sympathy he would bend the figure forwards toward the object of the emotion, with the fingers stretched in the same direction, as if ready to help, and the palm probably inclined downwards, as if in token of protection, but not because there is anything actually to be covered by them.

In exercising authority the body is raised to its full height, because the moral attitude is one of superiority, and the hand may be brought down to indicate that opposition will be dealt with in the way which in the symbolism of language is expressed as "put down." Again, a speaker in explaining his views may bring the fingers of one hand down on the other, as if he were producing a visible object and placing it on his hand before you, or were pointing to a visible statement on paper, the downward movement not now giving the idea of destruction, but of that which is symbolically called "laying down " his propositions. Here the movement of the hand keeps pace with the success of the speaker's effort to put his ideas in words ; the movement is arrested and the muscles tense, as in a state of mental tension he struggles with a difficulty; then as he overcomes the difficulty down goes the hand, as everybody knows, with energy parallel to that which he wishes to give to his statement.

It may be doubted if there are any gestures to which the principle of symbolism, which I have attempted to illustrate, does not apply. True, movement upwards, downwards, forwards, and backwards, with quickness or slowness, tension or relaxation, form but a small number of elements of gesture compared with the varieties of mental condition expressed, yet there are combinations, adjustments, and accessories which give them a wide range of expression. Take the condition of relaxation, for example ; it may be seen in enjoyed repose, or in sorrow, weariness, or despair. But, keeping feature entirely out of the question, in enjoyment of repose the attitude exhibits careful selection with a view to comfort, which in the excess of sorrow or despair is entirely absent. In despair the body is thrown back, and causes its relaxation to indicate the uselessness of having anything further to do with that which is before the mind's eye ; in weariness it turns to one side as if change could alone give relief from utter lifelessness ; while in sorrow the body is folded on itself as if the heart would nurse its own bitterness, and yield to prostration, with the world shut out.

But especially is the direction of the eye the appropriate supplement of gesture, and its connection with this may fairly be taken into account before considering expression by the features.

An erect carriage may be given to the body by haughtiness, conceit, the exercise of authority, or the presence of ennobling thoughts ; but very different is the direction of the eye in these different circumstances. In haughtiness the upward head contrasts with the somewhat downward glance, indicating that it is the height pertaining to self which occupies the mind and which looks down on others; in conceit the straying of the eyes over the person, and the glancing about to take note of the effect on others, show how approbation is sought for ; in command the glance is direct, as of one who would bring his personality right into contact with those whom he would wield ; but in ennobling thought the eye, as well as the body, is turned upwards as if both were governed by a power above them.

In kneeling in worship the idea is that of humiliation before a superior Being, and if the eyes are directed upwards, it is because the mind naturally associates the rule of such a Being over us with a dwelling above us ; while if the hands are clasped or crossed on the breast, it is the natural conclusion of a motion of the arms towards one another as if in desire to receive. I venture to think that this is a more natural explanation of clasped hands than the idea of placing them in the hands of another in token of submission. But what I wish to attract attention to now is, that in humble attitudes the direction of the eyes is in harmony with the direction of the face; if the face be turned upwards the eyes look upwards; if downwards, the eyes look downwards; and the expression is very different both from the contemptuous effect of an upward face and downward eyes, and from the downward face and upward eyes with their many variations, seeming, with one exception, always to convey an expression into which a concealed advance enters as a necessary element. The culprit sheltering himself by a lie, who has not mastered the base art of concealing the concealment which he practises, hangs his head over his secret, while he steals upward glances to see the effect which he distrusts; and if suspicion enters more largely into his feeling, he does not face you, but stands sideways, and, looking obliquely, betrays by the want of harmony between eye and attitude the duplicity which is within. On the other hand there are expressions delightfully gay in which a slight bend of the neck is combined with an upward glance; yet they have the element of slyness entering into their humour, or that equally innocent and slight suggestion of a secret taking the form of a confidence which seems to say, "You and I understand."

There appears to me to be one exception to the rule that a downward face and upward eye give the idea of concealment ; and it is in mental absorption when the head happens to be bent forwards and the eyes staring into space. Yet it is an exception more apparent than real, the glance having less the appearance of proceeding from the face than of having quitted it altogether. What catches the eye most in such circumstances is the relaxation, the absence of expression, from the mind being too much occupied with its musing to devote attention to attitude or feature. The head only bends when the relaxation of the previous attitude allows it to fall forwards; and it falls as readily backwards when the attitude has been favourable to that movement. The eyes also are probably nearly in the position of muscular inaction. In the dead the position of the eyes is more turned upwards than they would be in looking directly forwards during life, and their strange stare seems to depend less on the perfect movelessness than on a slight divergence of their axes. Further, it is known that the condi-i on of rest of the adjustments within the eyeball is when the focus is set for the infinitely distant, which requires the axes of the eyes to be parallel ;

and one may see in the stare of absorption that the eyes are parallel or slightly divergent, therefore probably with the muscles of the eyeballs relaxed.

But enough, probably, has been said to illustrate the principle sought to be established, that attitudes and gestures, including movement of the eyes, have direction corresponding essentially with the emotions which they express.

3. The same principle is applicable to the expressions of the features. It is palpable that in feelings of elation the angles of the mouth are raised, the upper eyelid also is drawn well up, the eyebrows are lifted, though not sufficiently to produce the slightest wrinkle of the brow, and even the lower eyelid is raised, partly by contraction of fibres of the *orbicularis*, partly pushed by the rising cheek. Nor is elevation the only movement, but nature expresses the expansive feeling, the tendency of gladness to widen its scope, by an outward movement. The angles of the mouth spread more outwards than upwards, and as elation is carried further the mouth begins to open. The apertures of the eyes are not as capable as the mouth of outward enlargement ; but to them also the appearance of greater breadth is given by the formation of lines spreading outwards and upwards from the outer angles. Nor is the nose, though less move-

able, quite quiescent : it is perceptibly broadened and shortened by the outward and upward movement of the alæ. Thus this characteristically human expression is not confined to a single feature, much less the effect of a single muscle, after the fashion in which Duchenne[1] endeavours to demonstrate that various expressions are produced each by a particular band of fibres. The means are various by which the results so harmonious in character, so similar in their symbolism, are achieved : here it is by the direct pulling of muscles, there by the accident of one part pushing another upwards, and by the wrinkling caused by an action having another primary object.

Under the influence of the depressing emotions the same parts are depressed which were raised in smiling ; and the apertures of the face, the openings of communication with the world, are diminished as the soul retires from its disagreeable surroundings. The brows, the eyelids, the alæ of the nose, but most of all the angles of the mouth, are lowered in all expressions of sadness.

But, in speaking of smiling and sorrow, I keep out of consideration altogether laughter, sobbing and crying. They do not fall within the limits of actions principally explained by the natural corre-

[1] Méchanisme de la Physiognomie Humaine.

lation of physical movements and mental actions ;
they depend, probably, altogether on the mechanism
of the nervous system. The nervous centres acted
on by an excess of emotion are deranged ; and it
would require a far more intimate knowledge of
cerebral function than has yet been arrived at to
enable us to follow the details of the causation of
the convulsions produced. Laughter, sobbing, and
crying have the feature in common of convulsive
breathing. In laughter, perhaps in symbolical connection with desire for the outflow of emotion, the
expirations are accentuated and prolonged, and are
therefore most obviously broken with convulsive
quiverings ; in sobbing, on the contrary, the inspirations are elongated and broken into a number of
convulsive acts ; while in the crying of children the
true sobbing is mixed with a desire to announce
their sufferings loudly abroad, and, therefore, the
convulsive inspirations are followed by an unneces-
sarily long expiration, utilised, if I may use the
expression, for the purpose of howling. But it is
interesting to note that the extreme distortion of
the face in the most violent crying is not dissimilar
from that in the most violent fit of laughter ; and an
amusing illustration of this can be obtained by turning up Plate I. fig. 2 of Mr. Darwin's work on
Expression, which represents a little child with eyes

shut and mouth open, evidently roaring. Cover
with a card all but the face, and draw on the card
the body of a fat old man lying back in his chair,
and the child's face, without a stroke of change, will
be converted into the bald head of the old man con-
vulsed with laughter. How so? Simply because
old men are more given to roar with laughter than
to bellow like children.

There is another of Mr. Darwin's illustrations with
which a similar experiment may be made, namely,
Plate V. fig. 1, a female head expressing disdain.
Hide the neck, and make that head bend over a
figure so drawn that the head shall have a droop in
keeping with the direction of the eyes, and the expression of contempt completely disappears, giving
place to one which is serious and quiet. The experiment, however, could not have been successful
if the expression had been carried further by the
curling of the upper lip. When contempt is expressed merely by attitude, it is done, as I have
said, by upward and backward motion of the head,
and a glance in precisely the opposite direction.
When the features aid the expression, they act on
the same principle. While the angles of the mouth
are free from all elevation, or are even depressed, in
token of the depressing effect of the unpleasant, retreat upwards and backwards from that which excites

disdain is indicated by the raising of a portion of
the upper lip; and the expression once originated
can be exaggerated by the drawing up of the lower
lip and the chin by the *levator menti*, while the
angles of the mouth are actually pulled down, so as
to give it the appearance of being held down at the
ends while an effort is made to pull it up from
the surroundings which hold it.

In the allied expression of disgust, the *levator
menti* takes no part, while the depressors of the
lower lip are more strongly contracted, because
the idea is no longer to keep away from the
objectionable notion, but to get rid of the foul
thing which has already entered. The same
muscles come into play in getting rid of a bad
taste; and language, travelling in a similar line
to expression by feature, signifies the alliance
by the word disgust. Darwin also, quoting Piderit,
draws attention to the action of the nose, which
gives the idea of getting rid of an offensive odour;
but I think these writers are mistaken in imputing
to the upper lip an action "so as to close the
nostrils as by a valve." The upper lip is incapable
of shutting the nostrils, and is not used in any
animal for that purpose; and what we really do, in
an unrestrained expression of disgust, is to raise and
distend the nostrils, as if to give egress to an objec-

tionable vapour which has already intruded; and forthwith we expire through both nostril and mouth.

Resolution is expressed mainly by the mouth; compressed lips indicating opposition to assault or entreaty from without, or the closing up of any tendency to yield from within. In the pout of discontent the angles of the mouth are drawn in as in all expressions of dissent, while the projection of the lips is in readiness for an expressive expiration rendered audible, it may be, by a labial; just as, when the teeth and jaws take up the symbolic action instead of the lips, naturally syllables with dentals lie to hand. By what slight circumstances expression of features may be modified is seen by the very different impression produced by a protruded and contracted mouth ready for an inspiration.

Hitherto I have made little mention of the muscles of the forehead. What Mr. Darwin has written on the "grief muscles"[1] is exceedingly suggestive; yet I own that, to my mind, it falls short of explaining the phenomena which he so graphically depicts. The frown expresses displeasure by the descent and gathering together of the two eyebrows, and so far is in perfect keeping with the principle of symbolic correlation which I have tried to give

[1] Op. cit. p. 181.

prominence to. It occurs also in attempts to see distinctly external objects, helping to shade from dazzling surroundings, and concentrates the attention on a limited field ; and, in exercising the internal perceptions when a difficulty is encountered, the forehead falls into the same condition as when a difficulty in distinguishing an object with the eye is met with. These explanations appear to me more satisfactory than to suppose that the frown is the relic of childhood's screaming,[1] the more so as in the worst cases of screaming in childhood the eyes are not protected by a frown at all, but by the violent closure of the lids by means of the *orbicularis*, so that the infantile frown is rather derived from the same source as the frown of the adult than the parent of it.

The formation of "rectangular furrows" on the forehead, with elevation of the inner part of the eyebrow, is more complex, and demands reference to the anatomy of the muscles. Undoubtedly we owe to Duchenne the knowledge, which anatomists ought to have perceived before, but did not, that the *pyramidalis nasi* is antagonistic to the central fibres of the *frontalis*. It is attached to bone below, while the more fixed attachment of the *frontalis* muscle is above. For a much longer time we have known the

[1] Darwin, op. cit. p. 225.

corrugator supercilii as an opponent of the *frontalis*, but the description of it is unsatisfactory. In the eighth edition of Quain's Anatomy it is described, according to the received mode, as proceeding "outwards and a little upwards." Luschka describes it as a part of the *orbicularis*, and denies its power to corrugate the eyebrow.[1] Henle, highly elaborate, describes it as part of the *orbicularis*, and consisting " of two or three slips covering one another in such a way that the higher they arise they are the deeper, and pass more from a gentle upward slope to a transverse direction."[2] But he also describes as part of the *occipito-frontalis*[3] slips passing upwards from an attachment to the frontal process of the superior maxillary bone. So far as I can see by dissecting the *frontalis, pyra-midalis*, and *corrugator* muscles from the deep aspect, a much simpler description would be more accurate, as well as more in accord with what may be made out by examining carefully the movements of the integument during life. I am disposed to describe together the muscular fibres passing upwards from the superior maxillary bone and inner end of the superciliary ridge as a sheet which widens as it passes upwards inseparably connected with the *frontalis*, its

---

[1] Luschka, Anatomie des Menschen, iii. p. 365.
[2] Henle, Anatomie des Menschen, Muskellehre, p. 143.
[3] Ibid. p. 136.

upper and inner fibres directed toward the frontal eminence, while its outer fibres form the part of the muscle usually bearing the name. In antique statuary the line of action of the moveable attachment of this muscle is sometimes indicated by a depression rising upwards and outwards high on the forehead, while the gathered integument is comparatively smooth. This is the actual line to which the action of the muscle can be traced on the living body in males with fleshy foreheads; and I have no doubt that this, unmixed with "rectangular furrows," was the outline of the horse-shoe of Sir Walter Scott in *Redgauntlet*, to which Mr. Darwin refers. When this line is drawn downwards and inwards, the result will be to approach the upper attachments of the outer fibres of the *frontalis* so much to their inferior attachments that they will be deprived of the power of raising the eyebrow or wrinkling the forehead; and this I believe to be the anatomical reason why the outer parts of the eyebrows are depressed while their inner ends are elevated in the joint contraction of *frontalis* and *corrugator* muscles. On the other hand, the outer angles distinctly participate in the action when the brows are raised in expression of cheerful eagerness; and when I assume that expression I can feel that the *occi-*

E

pitalis and *attrahens auriculam* are brought into action.

The raising of the inner ends of the eyebrows, while they are depressed in the rest of their extent, is not a downright and intense expression of one mastering emotion, but comes from a struggle of conflicting feelings. It is a combination similar to the raising of the upper lip while the angles of the mouth are drawn down ; and it often accompanies that action in expressing the acuteness of the petty vexation which is akin to disgust and sometimes so named. Then the expression travels up from the mouth to the forehead.

This brings to our notice that some expressions are more liable to be shown by the mouth, others by the eye and forehead, and only when they become intense do they invade the whole face. I believe that I shall be correct in saying that expression for the information of others is most liable to be made with the mouth, the organ of communication with the world ; while expressions that betray thoughts unintentionally with the outer world are most liable to begin in the eye and forehead.

I take it that the transverse wrinkling of the brow is simply the expression of mental irritation by muscular contraction, such as occurs in the rest of the face, but confined to the forehead when the mind is thrown in on itself and not intensely excited. Thus it is proverbially the expression of care, and still more of despair. Round both the mouth and the eyes, the muscles expressive of control are those which draw the parts together. When control is lost altogether, the radiating muscles have it all their own way ; and therefore it is that the brow is transversely furrowed, the eyes staring, and the mouth wide open in terror. When control is sought to be exercised, and emotion pulls the antagonist muscles, the result is quivering, a quivering sometimes seen in the rectangular furrows on the forehead. Those furrows are most variable in meaning. In bright sunlight the frontal is employed by the will against the instinctive protective action of the corrugators. At other times the frontal takes the lead, and the corrugators try to counteract it in an attempt to look grave, or an absurd trifle disturbs serenity, and twitchings take place out of harmony.

But I have sufficiently spoken of the subject which I have mainly sought to illustrate, namely, that the principal key to a great part of expression is the correlation of movements and positions with ideas. I shall only add that the correlation which I have sought to make plain is found elsewhere in

nature besides the face of man. In the vegetable
kingdom the flower is put in the place of honour;
in the vertebrate animals, the nervous system,
which, with the exception of the supracœsophageal
ganglion, had been inferior in the articulata, becomes superior; while in man the brain is superior
in every sense.

III.

VISION.

III.

VISION.[1]

FEW truths can be more surprising when first we become aware of them than the indirectness of the communication of the mind with the external world through the senses. Our sensations, like our movements, are the results of action of most complex machinery working without our having knowledge of its structure, or even, it may be, of its existence. We have no consciousness of the myriad changes taking place within it before a momentary sensation of sight, or sound, or touch can occur.

Not only so, but it is impossible to give any reason why the peculiar actions of the various sensory mechanisms are followed by the particular sensation which they each induce. Thus, although science discovers to us the existence of vibrations of light and sound, we cannot say why the irritation of

[1] In greater part re-arranged from portions of two popular lectures.

appropriate organs by those vibrations, should produce, one the sensation called sound, and the other
a sensation altogether incomparable with it, namely
vision. Neither can we say how it is that the vibrations of heat produce a sensation different from
either. In short, beyond the sequence made known
by experience, we know of no relationship between
sonorous, luminous, and thermic vibrations on the
one hand, and the sensations of sound, sight, and
heat on the other. The sensations may be purely
arbitrary consequences of the irritations produced
by the various stimuli, for all that science can tell
us to the contrary. We perceive in this an instance
of the impossibility, in our circumstances, of bridging
the gulf between matter and consciousness.

While, however, the sensations received through
the eye, the ear, and the surface of the body, so
differ in kind as to be incomparable, they yet
combine, by means of the addition of time and
space, to give common information. It is admitted that time is an idea independent of the
external senses; and, notwithstanding opinions
to the contrary, the position is possible that
the same is true of space, and that the mind
must from the first recognise itself as surrounded by *non ego* as well as existent. However that may be, a sense of the position of the

parts of the surface of the body is counted as a part of common sensation, though a very different thing from touch ; and it is from this that the first conceptions of magnitudes, including distance, are derived ; while we afterwards learn by experience to translate certain phenomena of sight and sound as effects of distance. As to tastes and smells, they differ from objects of sight and hearing in being referred to the situation of their respective organs ; yet they may, when accompanied with movement, and the comparison of differing degrees of the sensation taking place one after the other, become associated with the appreciation of locality, as occurs remarkably in dogs following the scent. And not only is the exercise of the senses thus mixed with the idea of space, but every sensation involves consciousness of its duration and repetition or non-repetition. Thus the ideas of time and space become the means of unifying the results of sensations incomparable in their own nature ; so that the hand, the eye, and the ear, combine to increase the common stock of information.

It follows that extension, configuration and movement are properties of external objects which fall under a different head from colour and sound as we see and hear them. A sense of redness has no objective reality, but the vibration which causes it, the repeated movement at a definite rate, is in a very different position, and has exactly the same objective reality as space and time, whatever that may be.

In examining the characters of the channels of communication between the mind and the world afforded by the senses it will be convenient to consider for a moment the mechanism of touch before passing on to vision.

Suppose that you put your finger in contact with the point of a pin, and let us try to make out as far as we can the mechanism by which you become aware of that very simple matter. In your finger there are numbers of threads, each branching at the end into a brush of filaments so densely distributed that you cannot place the pin point on any part of the skin where it will not be microscopically close to some of them. These threads, called nerves, run up to the spinal marrow ; and when your finger touches the pin it is most certain that a change takes place in the end of a nerve and quickly travels along it, at a rate of more than one hundred feet per second, till it reaches the spinal marrow. And even then you do not feel the pin-point; for if your spinal marrow were divided at the top of the neck, as has happened to some unhappy persons, we might pinch and burn your hand, and, though it would wince and jerk so as to give the appearance of feeling, you would be quite insensible to every injury of it. The change which began in the nerve must run up the spinal marrow and into the brain before you can feel the pin-point. And although science can tell you that when an animal's brain is removed it loses consciousness, and can tell you that intelligence is in proportion to complexity of brain structure, science cannot tell you why it should be so, nor why any commotion of particles of matter, the only change of which brain is capable, should be associated with operations of the mind, and even with the knowledge that the pin-point touches your finger. Thus we see that there is not only a long and difficult journey between the touch with the pin and our knowledge of it, but that there is a gap between our conceptions of mind and matter, utterly bridgeless in our present state of being; although all our ideas are shaped from that world of matter, our connection with which is so incomprehensible. Let us even believe, as I do, that the mind reaches more or less thoroughly at different times to the finger ends, and is actually present there so long as the connection with the brain is undivided, we do not by that supposition bring the mind into immediate contact with the external world; for you will admit that not one of you is conscious of the brain and nerves in which your consciousness resides, or would even know that you had such things unless those to whom the interior of the body is known by dissection had told you.

It is very startling, and to the unscientific very confusing. I do not care at the present moment to press the inquiry further. All the length which I wish to lead you just yet is to appreciate that even in touch, which is the simplest of our senses, there is no immediate contact of the mind with external objects. At most, you are only aware of the positions in which your various sensations are felt : you are not conscious of the things which cause those sensations.

Let us turn now from common sensation to vision. Vision is the most subtile and most complex of all the senses. It is produced by the vibrations of light, vibrations vastly more minute than those of sound ; and even sonorous vibrations are for the most part incapable of affecting the sense of touch ; although, no doubt, if one stands near a piano with a tense parchment in his hand, he will feel the notes producing vibrations in his fingers. Further, we not only appreciate light and darkness by vision, but the exact relative position of very numerous spots of light, with their various degrees of intensity and their differences of colour, all at the one time.

Every picture consists of points of light of different degrees of intensity and varieties of colours, and the appearance of form is nothing but the grouping of points of light and shade and colour. That the mind, then, which is only affected by the external world through the medium of nerve-terminations, may appreciate the landscape, it is necessary that it shall receive a separate sensation from the light emanating from each of a large number of points, and that the relative positions of these sensations shall be such that they may be recognized in positions corresponding with the points in the landscape to be represented.

In the eye these requirements are furnished. In it a camera or dark chamber of notable size exists similar to that which a photographer uses, having a lens in the fore part, and a sensitive curtain at the back. All the difference between this camera and a photographer's is, that this is globular and the optical arrangements are more perfect. When the photographer looks in at the back of his camera, he sees on the ground glass plate the image depicted which he wishes to photograph, placed upside down, but faithfully delineated in all its colours ; and such an inverted landscape is formed in like manner in the back part of each of our eyeballs. And as the photographer adjusts the

focus of his instrument by altering the position of
the lens, screwing it nearer or further from the
screen, so we adjust the focus of our eyes instinc-
tively according to the distance of the object
looked at, not indeed by changing the position of
the lens but by altering its form so as to make it
stronger or weaker as required.

The sensitive curtain is called the retina ; and
in the early embryo, when the brain is as yet
a hollow structure with delicate walls, the retina
was continuous with the brain, forming in fact
a portion of it, and lined on its deep surface
with a layer of cells continuous with the epithelial
cells lining the hollow of the brain. These cells
at the back of the retina become developed into
elongated structures called rods and cones, or
included under the common name of bacillary
elements ; and the exceedingly minute bacillary
elements are the organs on which the light of the
inverted picture acts. They have a highly complex
structure which need not engage our attention at
present ; and the main part of the retina presents
a remarkable modification of brain-structure, receiving the fibres of the optic nerve and placed in
structural continuity by means of other nerve-fibres
with the bacillary elements. The rays of light
pass through the retina proper without producing

any irritation, and reaching the bacillary elements stimulate the nerve - terminations in connection with them ; and when the impressed conditions thus started are continued through the retina and optic nerve to the brain the sensations of vision take place. Thus the mechanism of vision, though more complex than that of common sensation, is of a very similar description.

There is, however, one striking distinction. In the case of touch the sensation is referred to the spot where the stimulus is applied : in the case of vision it is referred, not to the spot where the ray of light falls on the inverted picture in the back of the eye, but to a point outside the body and placed in the direction from which the ray has come. That is certainly the case, however inexplicable ; although it is true that the recognition of distance is learned slowly by experience. Were the landscape at birth perceived as a pair of minute inverted pictures lying inside the head, it would be impossible for a new born animal to learn that they were indices of a world outside : and it is curious that physiological writers have not always noticed this, but have tried to explain how two inverted pictures, spoken of as if they were outside the body, are rectified and made one.

The perfection of vision necessary for the appreciation of a picture does not appear at once in the simpler forms of eyes met with in animals low in the scale ; and perhaps the most interesting way of looking at the structure of the human eye is to compare it with others of a less complex sort. The simplest form of indubitable eye is a nerve-termination with pigment on it, and more simple still there are spots of pigment in various animals in which nerves are not to be found; and those spots are possibly or in some instances even pro-bably eyes ; that is to say sensitive to the rays which the pigment reflects.

The star fishes furnish examples of a simple form of indubitable eyes. They have not all got eyes ; but those which possess them have them at the tips of their arms in groups which form a scarlet spot. We know that these organs are eyes because they have not only bright colour round them, but have a transparent structure at the extremity, which admits light to the nerve-extremities behind it. There can be no doubt, therefore, that light is the irritant which acts on these organs ; but it is equally certain that they are utterly incapable of displaying the forms of surrounding objects, for every ray which enters must fall on the red cup which surrounds the nerve terminations, and in consequence of this its red elements will be reflected from place to place in the hollow cup, and strike the nerve extremities in different places, while the other colours are absorbed. Therefore only red rays can be appreciated, and even these cannot be discerned as distinctively red, since there are no other colours perceived with which they might compare. Such eyes may guide the animal to localities where there are conditions of light favourable to its wants, just as a dog is guided by smell ; but it is difficult to imagine that they can give information more precise. Nor is it easy to see that much advance can be gained in function even where a lenticular transparent body lies in front of a number of distinct nerve-terminations, if as in the case of the emerald eyes that adorn the pecten, a brilliant pigment surrounds the whole. But when a transparent structure with nerve-termination behind is surrounded by dark pigment, which absorbs the oblique rays and allows only the direct rays to penetrate to the nerve, the effect is very different, and complication suitable for the production of vision may occur either by the crowding together of a number of such organs, or by the multiplication of nerve-terminations of a bacillary character behind one common lens in a dark chamber.

Of the first sort are the compound eyes met with in crustaceans and insects. If you look, for example, at the large domes which form the greater part of

F

the head of a dragon-fly you will see that even to
the naked eye they exhibit a grated appearance, such
as might be presented by a very fine sieve ; and under a magnifier they show a surface like that of a
honey comb, consisting of numbers of hexagonal
compartments. Every compartment is the end of a
tube, containing a transparent substance like glass,
its walls lined with black, and its deep extremity
occupied with a sensory structure continuous with
the nerve of sight. The effect of this arrangement
is obvious ; only one ray of light can pass down each
tube ; all lateral rays being absorbed by the black
walls. The sensitive structure at the bottom of each
tube is thus exposed to only one spot in the landscape, the spot directly opposite it; and there is
painted in the bottom of the eye a miniature repetition of the landscape which might be compared with
a pattern in Berlin wool, each tube containing a single
stitch. Here then you have undoubtedly an instance
of true vision—a number of separate spots of light
appreciated at the same moment, and every spot referred, not to its position in the eye, but to the direction from which the ray of light has come. This
sort of eye is an optical instrument which requires no
focusing. There are just as many points in the
landscape exhibited to the animal as there are tubes
in the eyes, and if the landscape is distant those

points are scattered, while if it is near they are crowded together; and probably in this way a sense of distance is obtained. We may judge also that this form of eye is suitable principally for seeing very near objects.

In all the higher kinds of animals—namely, in all the vertebrates, from fishes up to man, and likewise in the highest group of invertebrates, including the nautilus and cuttle fishes, the arrangement exists of a camera, a lens, a bacillary layer and a retina or membranous expansion of brain-matter. So far, the eye of the cuttle fish, together with the other less developed camerate eyes found in the invertebrata is similar to the eye of vertebrate animals ; but while the optical contrivances in the two sets of structures are similar, the sources of their origin are totally different ; so that it is impossible to conceive that by any process of modification in successive ages the one kind of eye could have grown out of the other. This is particularly the case as regards the retina or sensitive curtain on which the light is thrown. In both vertebrate and cuttle fish eye, it consists of a sheet of nervous substance connected with a covering of microscopically minute rods which receive the rays of light and are affected by them. In both instances it is an inverted picture which is cast on the retina, not an erect picture

as in insects ; and in both instances each rod receives its own ray of light which produces a separate
sensation, such as I have presumed to be produced
at the bottom of each tube of the insect's eye, and
in both the number of those separate sensitive points
is enormously greater than in the insect's eye.
To illustrate the difference in sensitiveness of the
higher forms of eyes and the insect's eye, I may
mention that in a fly's eye there are about 8,000
tubes, whereas in the most sensitive spot of the
human eye there are probably about 700,000 sensitive points crowded into about the two hundreth
of a square inch. But the resemblance between
the retina of the cuttle fish and that of vertebrates is carried no farther than I have mentioned. The rods of such a retina as ours are
structures which originated, as already mentioned,
in the earliest development, from the lining of
the interior of the brain ; while those of the invertebrata are derived from the skin ; and in consequence of that, there is this great difference, that in
the invertebrata the rods are on the surface of the
retina looking towards the light, while in our eye
they are turned away from it.

There is a way in which those two forms of eyes
may be more closely compared morphologically ;
and it is one which gives promise of more beautiful

harmonies to be yet discovered, exhibiting an orderly evolution of structure in the animal kingdom; but it is difficult to see what help it can give to those who must needs have the evolutions of organization accounted for by natural selection. In the very early vertebrate embryo the hollow of the interior of the brain was an open groove, and thus the epithelium lining it, including the bacillary layer of the retina, is originally continuous with the cells of the cuticle all over the body. The lens in both vertebrates and invertebrates is cuticular growth; and thus in a sense the whole eye is a superficial development in both.

But the distinction remains that the vertebrate eye is derived from two separate hollows placed apex to apex, and one folded round the other, while the invertebrate eye represents only one of them. Having regard to certain transparent tunicates and to Kowalevsky's observations on the development of the ascidians, in which the eye takes origin from the walls of an originally open neural sac, it seems even possible that the cerebral part of the vertebrate eye is the part homologous with the invertebrate eye, while it is the vertebrate lens which is the superadded structure. But such a suggestion is little more than a speculation. What is important is this, that the lowest vertebrate has no eyes properly

so called at all; that the animals supposed to be
its nearest invertebrate allies have only eyes of an
imperfect description incapable of true vision, while
other vertebrates have eyes which repeat the landscape, and are only comparable with invertebrate
eyes in the manner mentioned.

But however interesting the evolutions of the
organ of vision may be, it may be questioned if
there is not even more food for reflection in the
very existence of such a sense as vision or such
a sense as sound.   One has no difficulty in agree-
ing with Gegenbaur that certain eyes may have
been evolved out of tactile organs, but it is much
harder to understand how it comes to pass that
touch is the sense which it is, or that there can
possibly be any transition from touch to hearing,
or from touch to vision.   As sensations they are,
as already said, incomparable, and intermediates
between them are therefore inconceivable to us.
No doubt the sensations of sound and vision are
themselves inconceivable to those who are born
deaf and blind, but that very consideration points
to this: that in these senses there are two things for
consideration, the material apparatus and the properties of that consciousness which is capable of
being modified by stimuli received through the
medium of the apparatus.   To me it appears plain

that the idea of vision must have previously existed before it could form part of the consciousness of any animal; and in the evolutions of organs of sight I am compelled to recognize in the simpler forms the early stages of a morphological design, moving forward in definite directions to accomplish a mode of contact between the external world and the consciousness of animals, the idea of which already existed.

Is it in accordance with anything that we know of the laws of nature that such contact should be of an arbitrary and purely artificial kind ? Consider that while we are without experience of spirit-life, except in connection with body, it would be credulity to suppose that no spirits exist save those enchained by matter. Suppose them to exist, and suppose them to appreciate the material universe in that intrinsic character which reason and not sense informs us of, namely, as so many centres of force inhabiting space, it is plain that they cannot have vision in the sense in which we have it—a sensation artificially produced through affected nerves. They cannot have any of our senses. But must they necessarily be devoid of the ideas which they represent ? Is it necessary to suppose that things which we possess in common with the majority of animals exist nowhere else in the whole universe ? Is it philosophical to suppose that there is no other universe than that which is indicated to us by whirling spheres that roll through space, and that these must be shorn even of those properties by which they directly affect our senses, the properties by which alone we have acquired knowledge of their existence ? It is easy to see what strange possibilities may be opened up by such a question, but that is no reason why the question should not be asked. Either we must suppose that the sensations of sight and sound have links in some way corresponding to them in a world of which we know nothing, or that they are arbitrary illusions, graduated doubtless in keeping with movements not directly appreciable by us, but arbitrary nevertheless.

IV.

THE PHYSICAL RELATIONS OF CONSCIOUSNESS AND THE SEAT OF SENSATION, A THEORY PROPOSED.

IV.

THE PHYSICAL RELATIONS OF CON-
SCIOUSNESS AND THE SEAT OF
SENSATION, A THEORY PROPOSED.[1]

IN venturing to disturb the theory of Sensation as it has long been taught, I am very sensible that it may be difficult to obtain a patient hearing, seeing that it is a theory universally received ; yet there are important points which that theory leaves unexplained, points familiar to all, and which only require to be mentioned for every one to admit that, as the theory at present stands, they are totally inexplicable. That theory may be shortly stated thus :—that an irritation applied in the neighbour-hood of a nerve-extremity produces an impression which is conducted along the nerve till it reaches the seat of consciousness in the brain ; and that the mind, affected by the impression, becomes thereby cognisant of a sensation, which it refers to the

[1] Originally published in the Journal of Anatomy and Physiology. November 1870.

extremity of the nerve along which the impression has been conducted. Further, let it be recollected that nervous impression is nothing but a physical condition, some of the peculiarities of which have been laid bare by experimenters, and which is capable of affecting any of the nervous elements, viz., nerves, both motor and sensory, and nerve corpuscles.

In speaking of nervous impression thus defined, we deal with a matter of fact, although we are not thoroughly acquainted with its details ; but in stating the doctrine of the modus operandi of sensation, we have merely to do with a theory.

This theory is, however, a physiological as well as psychological theory, and involves the consideration of the functions of nervous structures, as well as the laws to which consciousness is subject ; and it is the more important to point this out, because the physiologist is liable to think that where consciousness is involved physiology cannot be concerned, whereas the doctrine of sensation, although it relates to a matter on the psychological frontier, is arrived at from physiological data ; and it is because it is, as will be shewn, at variance with other anatomical and physiological data that it requires alteration.

It may be taken as certain that when a nervous

impression is conducted to the seat of consciousness
in the cerebral hemispheres, corpuscles there enter
into a corresponding condition to that of the conducting nerve, precisely as a corpuscle which is the
turning-point in a reflex action is no doubt affected
by the condition which it receives from the sensory
and passes on to the motor nerve. But if we
imagine that in an act of consciousness the corpuscles of the hemispheres undergo any change
other than that of passing into the impressed condition studied in nerves, we become guilty of an
assumption which has the plain objection of being
unfounded. That assumption is unfortunately
often made, apparently from confusion of ideas ;
for authors, particularly medical authors, speak as
if mental impressions lodged in the brain ; whereas
nervous impression, the only active condition into
which there is evidence of the brain passing, is
a physical state of a uniform nature, while, on the
other hand, a mental impression is the presence of
a notion in the mind, and the variety of such
notions is infinite. I shall return to the consideration of the bearing of this remark on views as to
the details of the functions of the hemispheres, but
wish first to direct attention to the difficulties of
the received doctrine of sensation.

I own that, even supposing all difficulties as to

the route pursued by impressions from the periphery to the brain to be removed, to me it is
utterly inconceivable that the sites of irritations
over the whole surface of the body should be
minutely indicated, and a more vague appreciation
of the positions of internal irritations be obtained,
by differences in the cerebral termini of the impressions conveyed from different parts. The supposition involves the difficulty that there is no mode
by which the mind of the child could ever learn to
associate the changes taking place at the cerebral
termini with changes at different parts of the
surface. If the consciousness were ignorant of
the surface of the body at the commencement
of life, it must continue always to be so, for want
of means from which to derive the information ;
no amount of custom could avail it, for the mind
gets all its experience from the senses, and until it
could refer sensations to the parts of the body
from which they were derived, it could gain no experience of the external world whatever. Therefore
the doctrine of sensation necessitates the assumption that the functional union of the parts of the
· periphery with different termini in the brain is
primordial, and that the surface of the body is
minutely represented or repeated by a number
of points in the brain, which, however confusedly

massed together, derive their properties from their connections.

This assumption has possibly never been stated in this form before, but it is one from which the received doctrine of sensation allows no escape, and surely it is an assumption sufficient to arouse some suspicion of the doctrine which demands it. If it be true, it seems strange that some of those termini are not often the seat of disease while others escape, that we do not meet with paralysis of the sensation of limited patches of the body occurring from limited cerebral lesions, and that nothing of the sort has been laid bare by the experiments of vivisectors.

If we now contemplate the routes by which impressions from the periphery reach the brain, we meet with other difficulties. It is plain that if the mind is affected by the condition of cerebral elements as if they were situated at the parts from which they receive impressions, then the accuracy of the mind's knowledge of the periphery is dependent on the number of such elements which receive impressions from distinct parts, and that each distinctly recognizable spot of the body must be joined by a separate tract with its own cerebral terminus. That tract may be interrupted by corpuscles with branches in other directions, and must be so to account for reflex actions, but it cannot in any part of its course, pass through a fibre common to it and another spot of the periphery, without the consciousness being bereft of all means of distinguishing the one spot from the other. But all that we know of the structure of the cord notoriously contradicts the notion of such an arrangement—an arrangement which would involve the continual accumulation of additional sensory tracts in the cord from its lower to its upper extremity. Moreover, it appears certain from the experiments of Dr. Brown-Séquard that the greater part of the conducting material of the cord, its white matter, may be divided, and, provided the grey matter is left intact, sensation in the parts beyond the lesion remains unimpaired. Thus, the routes of communication between the brain and nearly the whole surface of the body are known to pass through the extremely limited area exhibited by a transverse section of the grey matter of the cord, and probably more than half that area is occupied with binding tissue which has nothing to do with nervous conduction, while of the nerve-fibres which traverse it, a number are motor, and according to the theory some must be devoted entirely to impressions from internal parts of the body, even though such impressions are principally patho-

logical. Can it be believed that the remaining
fibres traversing such a section approach in number
what would be necessary in allowing one nerve-
fibre for the area of distribution of each nerve-
fibre on the surface of the body ?

If it be attempted to escape from this difficulty
by attributing with Dr. Beale a function of importance to those numerous striæ or fibrillæ which that
writer, and after him, Max Schultze, have described
in nerve-corpuscles and axis-cylinders, the importance of a single nerve-fibre as a conductor may be
indefinitely multiplied, but the assumption is made
that those fibrillæ are really present during life,
and that each one is capable of maintaining the
active or the passive condition quite independent
of the others with which it is in close alliance ; an
assumption which must appear sufficiently great
to those who remember how little individuality
belongs to the much more easily demonstrated
fibrillæ of muscular fibre. Even then, the difficulty remains as much as ever ; for, if we suppose
that these fibrillæ are continued down all the poles,
and that the peripheral nerves possess them, as
well as others, we have indefinitely multiplied the
tracts in the periphery as well as in the cord, and
left the disproportion in numbers the same as ever;
while, if we deny fibrillæ to the peripheral nerves,

G

and suppose them to exist in the fibres of the cord alone, we assume that the work which requires a whole fibre in a peripheral nerve, can be done by one of a multitude of striæ in a fibre of the cord, which seems improbable.

If the difficulties which have been enumerated appear great, when the received theory is applied to the explanation of common sensation, they are still more striking when it is applied to vision. The theory demands that for every ray of light appreciated by the mind there must be a completely distinct communication from one of the rods or cones which constitute the individual sensory organs of the retina to a terminus in the brain, and that the impressed condition of every such terminus must be capable of creating in the mind a knowledge of the position of the point on which the ray falls as related to all the other impressed points of the retina.   That supposition involves all the difficulty which has been pointed out in the case of common sensation ; while the anatomy of the retina, even more distinctly than that of the spinal cord, contradicts the possibility of distinct communication between each of the immense number of peripheral nerve-terminations and the seat of consciousness.   Within the retina, the threads leading from the terminal rods and cones already enter into a complicated ganglion most developed in the part where vision is clearest, and it would be very hard to imagine that the fibres of the optic nerve emerging from the ganglionic corpuscles correspond individually with distinct rods or cones.

These being the objections to the received theory of sensation, they appear to me to warrant search for an escape from them, and I have been led, by the consideration of the properties of the living corpuscles of the body, and of what appear to me the established facts with regard to the actions of the cerebral hemispheres, to a hypothesis which I venture to put forward, believing it to furnish that escape, and to be in harmony with all that is known of the nervous system.

But before doing so, I find it necessary to explain and defend the view which I hold with regard to the connection between mind and brain, and which I have already in a few words suggested in a paper on the structure of the cerebral convolutions in the *Quarterly Journal of Microscopical Science* (April, 1870.) We must inquire into the relations of consciousness with the hemispheres before we ask how it is brought into communication with the finger-ends.

The cerebral hemispheres, which are well ascer-

tained to be the organs without which none of at
least the higher acts of consciousness can be performed, are a pair of highly corrugated vesicles,
closely connected one with the other, and continuous with the rest of the brain by the lower margins
of the inlets into the cavities of the vesicles, where
are situated the corpora striata. The structure
throughout the different lobes and convolutions of
the hemispheres, notwithstanding minor differences,
is fundamentally the same. Their grey matter consists of strata of nerve-corpuscles and nucleated protoplasm, and the vast majority of the corpuscles are
connected with fibres ascending from the corpora
striata and with others which lie horizontally, that
is, parallel to the surface, while they send another
stronger process perpendicularly outwards. The
horizontal fibres exist throughout the thickness of
the grey matter, but are massed particularly in
a varying number of strata, while others of them
coat the surface in various places. We have thus
an arrangement fitted to allow consentaneous
action of the whole grey matter of the hemisphere,
but nothing pointing to different functions of the
different regions. The convolutions are accounted
for when it is considered that they bring the
vascular supply within easy reach of every part
of the grey matter, and the minor differences of

structure may reasonably be supposed to arise from the different relations in which the different parts of the hemisphere lie to the corpora striata and from unequal development of the nerve-corpuscles ; a supposition in favour of which is the circumstance that in the posterior lobes, the parts furthest from the corpora striata, the nerve-corpuscles are least developed, and the horizontal fibres are most abundant.

But not only does the study of structure thus point to the probability of the whole hemispheres having one combined function, but development, comparative anatomy, and experiment point to the propriety of considering the corpora striata as forming with the hemispheres one organ. The corpora striata are enlargements of the basal portions of the hemisphere-vesicles ; so says development : they are inseparable from the islands of Reil ; so says adult structure. They may be irritated or damaged by vivisection in mammals, without damage to sensation or motion, and slicing them away in birds produces effects corresponding with those produced by slicing away the hemispheres in mammals ; that is the evidence of experiment : while comparative anatomy shows the hemisphere-vesicles forming each a unity in fishes, a small distinct corpus striatum within each in the turtle, and in the bird the whole hemisphere-vesicle converted into corpus striatum (or, more correctly, corpus striatum and island of Reil), with the exception of little more than a membrane at the upper part. The hemisphere-vesicle, therefore, appears to be a single organ primarily divisible, as Reichert has beautifully shown, into a root-part which includes corpus striatum and island of Reil, and the "mantle" which includes the remainder; and I apprehend that *the mantle is only a multiplier of the function of the root-part.*

If we now revert to one of the propositions made at starting, and bear in recollection that while there is every reason to believe that the corpuscles of the hemispheres pass into the impressed condition studied in nerves, there is no vestige of evidence that they have any additional active condition, it will become apparent that the law of operation of the functions of the hemispheres is that they are so connected with the mind that *the total amount of mental action at one time is dependent on the total amount in the hemispheres of that physical state which we call the impressed condition.*

Consciousness and the impressed condition of brain-substance go always together, but that impressed condition of the brain-substance is of one invariable nature, while the objects which may occupy the consciousness, its acts of memory, of observation, of reason, of volition, and its conditions of emotion or desire are of endless variety. The hemispheres, including the corpus striatum, may therefore not inappropriately be termed the organ of attention, the mode of whose action may be described thus:—the total amount of attention at any one moment is in proportion to the total amount of the impressed condition of the whole corpuscles of the hemispheres: but that attention may be occupied with numerous different mental actions going on at the same time; and with the specific nature of its occupation, whether memory, reason, emotion, appetite or volition, the brain can have nothing to do.

This view stands in direct opposition to the views which find favour at present with at least many physicians of mental disease. Starting with the proposition that the diseases of the mind are the diseases of the brain, which has a certain amount of truth in it, they have unconsciously slid from that starting-point into phrenological assumptions and into a belief that brain-corpuscles are pigeon-holes of ideas; and thus they predicate of matter properties incomprehensible and of which there is no evidence. But it is remarkable that neither medicine nor surgery, nor experiments on animals, have shown any difference of function in the different parts of the hemispheres. Wounds of the brain and limited spots of disease do not vary in their symptoms according to the part of the roof of the hemisphere attacked. I am not forgetful of the remarkable phenomena of aphasia nor the discoveries of Broca, but I have pointed out elsewhere[1] that the connection between aphasia and lesion of the brain external to the island of Reil is best explained, not by reversion to the doctrine of the divisibility of the hemispheres into different organs, but by taking into account the variety of processes required to go on consentaneously to produce speech, and considering the interruption to the consentaneous action of the hemispheres by interference with the fibres of communication at the most central part of the organ.

It is not only in the case of speech but in all the more complicated operations of thought that the mind has to attend to different matters at the same time, and it can well be imagined that simultaneous and harmonious action of the brain-

[1] Medical Press and Circular, 11th March, 1870.

elements is very necessary to secure that end. Thus it will be seen that the doctrine of brain-action which is here proposed by no means separates altogether the character of the mind from dependence on that of the brain. There are three elements on which the character of the brain as affecting the mind may be safely presumed to depend, viz., the intensity and ease of action possible to individual brain-elements, the total amount of those elements, and, lastly, but possibly, most important of all, the freedom and perfection of communication of those elements one with another. It may well be supposed that this last requisite is more liable to be deficient in large brains, as the distances are greater and the elements to be joined together more numerous in them.

In the structure of the brain there is the closest affinity to the structure of the rest of the nervous system, and its corpuscles are elements plainly comparable with the living parts or protoplasm-derivatives of other textures, and there is ground to presume that in like manner the impressed condition of nerve-corpuscles, whether in the brain or elsewhere, and of nerve-fibres is analogous to the contraction of muscular fibres and of amæboid corpuscles : the peculiarity of those of the hemispheres being simply that action of the mind excites their action, and that their action excites that of the mind.[1] That presumption prepares the way for the hypothesis of sensation which I venture to suggest, which is this : that *the consciousness extends from its special seat so far as there is continuity of the impressed condition ;* that when an irritation is applied to a nerve-extremity in a finger or else-where, the impression (or rather impressed con-dition) travels as is generally understood, but exists for at least a moment along the whole length of the nerve, and that as soon as there is continuity of the impressed condition from finger to brain the consciousness is in connection with the nerve and is directly aware of the irritation at the nerve-extremity. Or the position, may be shortly stated thus : functional continuity between nerve-extremity and brain is proved to be neces-sary for sensation, while on the other hand existence of distinct routes of communication between them is highly improbable; and seeing that functional continuity is sufficient of itself to explain the phenomena, we are not entitled to assume the

[1] A wider generalization may probably be made, namely, that every living element of texture exists in two conditions, the self-nutrient and the irritated ; and only in proportion as self-nutrition is in abeyance is the irritated condition possible.

existence of distinct routes, as has hitherto been done.

Let it be distinctly understood that I do not say that consciousness resides in the nerve-extremities, but only that when the nerves are in the active or impressed condition in their whole extent up to the brain, the consciousness is affected directly by the irritations applied to their extremities. The relation then of consciousness to the brain remains totally different from its relation to the nerve. The impressed condition of the cerebral corpuscles produces only excitement of the mind; the impressed condition of a peripheral nerve or its extremity continued up to the brain brings the consciousness into communication with the irritation applied.

At the extremities of the different nerves are placed various arrangements which are least complicated in the nerves of general sense, and which modify their capability of being irritated by different stimuli. Thus the expansion of the optic nerve is incapable of being affected by the irritation of light, but the rods and cones of the retina are irritated by them with ease. The complicated nerve-extremities of the ear are acted on by sounds, and the comparatively simple extremities of the nerves of general sense are acted on by mechanical irritations and changes of temperature. It appears then that if we suppose that the consciousness is directly affected by the application of the irritation to the nerve-extremity, we have the simplest possible explanation of differences of colour and of notes, and of the different kinds of sensation, conveyed by the nerves of general sense. We are in a position to say that as the irritation varies so varies the sensation, without being obliged to assume different kinds or conditions of nerves without evidence.

The objections may be urged against my theory that irritation of the ulnar nerve produces sensation in the finger ends, that persons after amputation of an arm feel pain in the fingers, and that irritation of one nerve often causes pain in another, as when disease of the hip gives pain in the knee, or toothache pain in the temple ; but it appears to me that all these things are better explained by this theory than by the received one. When one strikes the ulnar nerve at the elbow accidentally with force everybody knows that the acute pain is not felt in the fingers, but at the part struck, and that this is immediately succeeded by pain and a peculiar sense of vibration which travel downwards from the struck part till they reach the fingers. A patient who has had a limb removed, in like manner, is perfectly aware of the site of irritation when a nerve-trunk in the stump is touched, and it seems reasonable to suppose therefore that the sense of the presence of the removed part is some-how to be explained by habit, the reversion of consciousness to previous conditions, and is com-parable with other waking illusions of the phantom sort. The production of pain in the temple from toothache may be explained on the supposition that intensity of the impressed condition sufficient to produce pain is conducted to a nerve-centre and spread centrifugally, just as I have suggested that the alteration in an ulnar nerve struck at the elbow spreads down to the fingers.

If I have put forward these views of sensation simply as a suggestion I should be still more unwilling to be dogmatic in applying them to the motor nerves. Yet it may be mentioned that one of the most inscrutable of all phenomena, looked at from the received point of view, is the circumstance that the will is able to regulate delicately the movements of the limbs by adjusting the contrac-tions of complicated muscles of which the mind is wholly unconscious; and the difficulty in this matter will be greatly simplified if we can see our way to believe that, when a motor nerve is set in action from the brain, the mind is brought into

connection with its distribution. Further, the hypothesis facilitates the explanation of muscular sense,
and accounts for what has been observed by Dr.
Brown-Séquard, that the tracts for muscular sense
do not decussate in the spinal cord. The phenomena sought to be explained by the hypothesis of
muscular sense are the consciousness of the position of a part, the capability of regulating its
movements even when ordinary sensation is para-
lyzed, and the consciousness of the maintenance
of muscular effort in such parts. The hypothesis
which I offer in explanation of these phenomena is
that by continuity of the impressed condition from
the brain to the distribution of the motor nerves
we are conscious of the parts to which the distribution extends, and of the exercise of the will within
them. But, it will be alleged in reply, when we
move our fingers we feel the movement not in
the muscles of the forearm, but in the fingers. I
think, however, that this difficulty will not appear
so great when we analyse our sensations in endea-
vouring to make some unaccustomed movement.
Thus, although in moving our limbs we are not
conscious of any sense of effort which we could
localize in the region of the muscles brought into
action, unless perhaps when unaccustomed resistance is met with, it is different when one attempts

to move the pinna of the ear or bring the palmaris brevis into action. If I attempt either of these actions I am conscious of exercising an effort to produce contraction in the neighbourhood of the part, on that side of it towards which I wish to move it. It seems to me therefore that muscular sense, as I have sought to explain it, is sufficient, in conjunction with experience, to account even for the movement of the fingers.

The bearing of the hypothesis now put forward on the functions of the nerves may be expressed in a few propositions.

1. The irritation of a nerve of common sensation throws the nerve into the impressed condition; and as soon as that condition is continued to the brain, the mind recognizes the irritation at the site where it is applied, in the form of sense of touch, temperature, or pain, according to its character. Over-intensity of the impressed condition may also itself be recognized in the form of pain.

2. Nerves of special sense differ from those of common sensation both in the circumstance that the apparatus at their extremities is affected by irritations of a different kind from those which affect other nerves, and in their irritation being recognized in the form of the special sense to which they are devoted.

3. By the impressed condition continued from the brain to the distribution of a motor nerve not only is a stimulus communicated to the muscles and applied by the nerve, but muscular sense is given ; and the consciousness being brought into direct communication with the part by establishment of continuity of the impressed condition, the will is enabled to regulate the position of the part and the degree of muscular energy with which it is maintained. But a motor nerve differs from sensory nerves of all sorts in the fact that irritation of it does not produce any sensation either of the character of common sensation or special sense ; and in this respect it is probably like the proper fibres of the spinal cord and brain.

It may be allowable to say in conclusion that in publishing these views I put most importance on the objections which I have urged against the received doctrine of sensation ; and in venturing to suggest another in its place I am perfectly aware that there is much which is imperfect in the hypothesis which I have put forward, and I can only hope that it may prove of some use in building up a more perfect view.

V.

CELL THEORIES.

H

V.

CELL THEORIES.[1]

Stricker's article on "The General Characters of Cells," in his "Human and Comparative Histology." Published 1868; translated by Power, 1870.

Beale, in Todd, Bowman and Beale's "Physiological Anatomy and Physiology of Man," 1866 and 1871.

IN placing the titles of these books at the head of this article, I purpose less to review them than to refer to the opinions of their authors in illustration of a general survey of the conceptions at present prevalent, with regard to the vital units of living bodies.

Observation is more a means than an end. The end is to arrive at an accurate conception of the processes of nature; and very different conclusions are arrived at by different men, founding their

---

[1] Reprinted from the Quarterly Journal of Microscopical Science, July, 1873. The article, as it originally appeared, contained a criticism of the views held by the late Professor Hughes Bennett, but it has been deemed unnecessary to reprint that part at the present date. A few added sentences are placed within brackets.

judgment on similar phenomena. Possibly, the just zeal at the present day for accurate information tends to lead to an undervaluing of the faculty by which observations are translated. Every man is obliged to translate what he sees. No doubt he ought to distinguish carefully in his own mind the appearances seen from his translation of them ; but no man ever did or ever will give a description of a complex microscopic appearance so as exactly to reflect the appearances seen, uncoloured by the element of judgment ; and much less is it possible to found a statement on a variety of observations which is not largely dependent on the attitude of the observer's mind. One observation has its effect in modifying the translation of another ; and it is greatly due to this that many microscopic objects of a corpuscular nature, or what are still called cells, are capable of being interpreted very differently now as compared with the way in which they were looked at twenty years ago. Formerly, corpuscles round which no cell-wall was demonstrated were too easily supposed to have one although it was invisible, or were regarded as exceptions to a general rule ; now they are viewed with different eyes, and taken as proofs that the cell-wall is unimportant.

The history of conceptions regarding cells is in its general outlines exceedingly instructive. The notion of a cell was first derived from vegetable tissues with their easily exhibited cellulose cell-walls. The vesicular form thus caught the eye from the first. Then, in many instances in animal tissues also a real or apparent vesicular structure was easily observed ; as in adipose, epithelial, and nerve-corpuscles. The contents in vesicular structures were seen to be various, but the frequent existence of a nucleus of firmer consistence imbedded in them could be demonstrated, and within this were often seen one or more nucleoli of some sort or other. Thus it naturally happened that the cell-wall was considered a characteristic structure, and was supposed to be functionally important ; and next to it, the nucleus was regarded as the seat of vital properties, because it was seen to divide preparatory to the multiplication of cells, and to be distinct in young cells, however it might dwindle out of sight in the old. The circumstance, manifest from the outset, that cell contents were various, taken in conjunction with their being the part which least caught the eye, led to their vital importance being long overlooked. But as improved microscopic methods came into use, including reagents, such as carmine, which bring masses of albuminoid matter into view not only by outline but throughout their extent, the general existence of such masses in bodies of the sort which had been known as cells came to be recognized, and, as a natural consequence, it gradually dawned on the minds of independent observers that the outlines of such masses were the things which, in many instances, had been translated as cell-walls.

Two other advances aided in completing a revolution of opinion with regard to cell-walls, namely, the discovery of corpuscles undergoing amœboid changes of form and migrations, and the tracing of nerves in many instances into continuity with nucleated corpuscles. Thus a change has crept into the whole conception of the nature of those vital units whose importance functionally had been first recognized in the case of some which had apparently a cellular form ; and thus it happens that the term cell is still employed in many instances in which it would be better to use the word corpuscle. A nucleated corpuscle is as convenient an expression as a nucleated cell; a connective-tissue-corpuscle is an expression which involves no theory or description either of structure or function ; and it would be an enormous advantage to the spread of accurate ideas if the word cell were never used except when it was meant to predicate the existence of a cell-wall.

For some years past I have, in teaching, been particular in this matter of nomenclature, believing that misleading names do generate confused ideas, and that no conventional compact can make it judicious to designate a solid mass by a word which indicates a hollow vesicle, or advisable to use a common word in a sense at variance with its usual meaning. It is as if we were to invest the tongs with the scientific name of poker. "There is nothing, I can assure you, gentlemen," said Goodsir, "which has more retarded science and philosophy, and the kindred subjects on which human reason has been employed, than the introduction of terms with conventional meanings." But I admit that it is difficult to escape from an accustomed groove, and that for a time one must be content, under protest, to speak occasionally of secreting cells, nerve-cells, hepatic cells, and so forth ; were it for no other reason than to be in harmony with the language of text-books, in speaking to students.

The changes in the anatomical conception of the living corpuscle have not been without their influence on the physiological conception. In the days when the cell-wall was paramount, it seemed an important point to determine how far the passage of substances through that membrane was the result of mere osmosis, and how far it depended on the action of some attractive power within.　To the school whose tendency was to refer everything to the laws of dead matter, the cell-wall was a most important agent; to those on the other hand who considered that to account for the formative changes in living bodies the presence of another force must be assumed, the nucleus as situated in the interior seemed the source of vital actions.

Moreover, in the absence of a defined knowledge of the protoplasmic element, the conception of the nucleus was obscured by extending the designation to bodies which ought not to be so named.　An instance of this may be found in the case of connective-tissue-corpuscles.　Many years before Virchow's researches threw a light on these structures, a corpuscular element was recognized as present in connective tissue.　It was even taught by some to be the same element as could easily be demonstrated as constituting, in a cellular form, a large part of the bulk of fœtal connective tissue, and to be of the utmost importance in its vital properties.　But that element, in the adult, was known only as it may be found figured by Dr. Sharpey in 1848 (Quain's 'Anatomy,'

5th edition, pp. cxv. and cxvi.). No doubt Dr. Sharpey distinguished clearly the nuclei "attached to the surface of the filamentous bundles or in their interior" from the rounded and oval corpuscles and irregular particles met with in the interstices of the tissue," which he said were "probably to be considered as belonging to the interstitial fluid;" but those who imputed importance to that latter variety of corpuscles had little help for it, at that time, but to associate them with nuclei.

In the present day the protoplasmic element has assumed an enormous importance, casting the nucleus into the shade, while the reign of cell-walls has come to an end altogether. But to speak of life, as is sometimes done, as if it were an inherent property of a particular chemical substance, is surely going too far, and is a view which has nothing true in it which is not more than thirty years old; for it has long been familiar to every one that life never exists without the presence of nitrogenous substance of an albuminoid character; and, though it has since been discovered that life in various instances exists in non-nucleated structureless masses of protoplasm, that is a very different thing from life being a property of protoplasm. Further, it may very fairly be questioned if some of the simple organisms are not rather to be com-

pared with the nuclei in textures than with protoplasm around nuclei. If certain of them are non-
nucleated protoplasmic masses, may not vibriones
be regarded as mere nuclei and nothing else?
Besides, in the textures, there are many nuclei
which have no apparent protoplasm about them;
and there are also nuclei with processes, which may
be regarded as bodies intermediate in character
between the typical nucleus and the protoplasmic
mass. The corpuscles of the deep layers of the
cutis are mere nuclei, with long processes in various
directions; while in the tapetum of the eye of the
ox long threads extend from nuclei, like the threads
at the extremities of fusiform cells in fœtal connective tissue. Also, a spermatozoon may be regarded
as a simple nucleus. No doubt, as mentioned by
Stricker, both Schweiger Seidel, and la Valette St.
George declare that not only a nucleus, but protoplasm, enters into the construction of the spermatozoon; but if we examine the forms represented
by la Valette, both in " Stricker's Manual " and
in his original paper in " Schultze's Archiv," we
shall see that what is meant is that the protoplasm is at first adherent to the spermatozoon or
nucleus, and afterwards absorbed into it or otherwise lost, but that there is no permanence in
the spermatozoon of a substance preserving the

chareacters of protoplasm and distinct from the nucleus.

One of the short-comings of Professor Stricker's article on the general character of cells is that he exaggerates the virtue of the protoplasm. He uses the expression, " Protoplasm is termed a living substance," or, as the German (man bezeichnet) may be more strictly translated, "it is recognized as a living substance ;" and he speaks in such a way as to leave the impression that it is a definite chemical compound. Now, the fact is that protoplasm when examined under the microscope, is usually as thoroughly dead as anything could be well imagined to be. Living masses of protoplasm, no doubt, can be studied microscopically, and a great stride has been made in science by the examination of such masses in texture ; but the composition of the protoplasm is not definitely known. It is quite unobjectionable to call the albuminoid mass of a nucleated corpuscle protoplasm, even after it has been acted on by means of chromic acid, carmine, or other reagents. In fact, protoplasm is simply a convenient name to use in speaking of the pulpy nitrogenous substance of vital corpuscles ; but it is not to be forgotten that the substance referred to is variable in appearance and behaviour, as is well illustrated by Heidenhain's observations on the differences in both salivary and gastric secreting corpuscles in states of activity and rest. How, then, shall we say that in its different conditions the material which constitutes the mass of such corpuscles is one and the same chemical substance? We shall, indeed, take a very imperfect view of the living units to which an unhappy chance has given the unfortunate name of cells, if we say that because neither cell-wall nor nucleus is an essential element, therefore life is a property of protoplasm. It was recognized by observers long ago that the bond of connection between the bodies which they described lay not in a detail of structure, but in the possession of one or more of the vital properties, irritability, growth, or reproduction; and the observations of later years do not overthrow that conception, but afford it additional support. Indeed, there are passages in Professor Stricker's article which show an appreciation of this.

That article is one which affords much food for reflection, and is a repertory of important information. But as a history it is defective, even greatly so. The author neither does justice to the work of Virchow nor of Beale; and Goodsir is a name of which he makes no mention. Yet Virchow inaugurated an era in the history of cell-conceptions ; and Virchow dedicates his "Cellular Pathology" to Goodsir, as "one of the earliest and most acute observers of cell life." And if I may again quote one neglected anatomist in support of the claims of another, it may be mentioned that in 1845, Goodsir, in his paper on "Centres of Nutrition," which never admits the possibility of cells originating otherwise than from pre-existing cells, declares that "for the first consistent account of the development of cells from a parent centre we are indebted to the researches of Martin Barry." However, we are informed by the German Professor, that Remak, in his "Entwicklungsgeschichte," 1852-1855, "has the merit of chiefly contributing to the abandonment of the doctrine of cell formation from free blastema," and that the same observer established the law that cells are developed by division alone in pathological processes also. Then follows the remark, naive enough, considering these statements about Remak, that Virchow's "well-grounded statement made in 1855, 'Omnis cellula e cellula,' really constitutes the basis of our present cell theory." Virchow's real claim to consideration in the history of cell theories is neither grasped nor mentioned, namely, that by displaying in their full importance the connective-tissue-corpuscles, he afforded the means of accounting for the development of the more complex elements of tissue, as well as for pathological growths.

These omissions, however, are accounted for by a note inserted by Professor Stricker in the translation (p. 38). Misled by Cohnheim, he had believed the results of Goodsir, Redfern, and Virchow to be founded on incorrect investigation ; but later observation of his own has convinced him that he was mistaken. These are not his words, but perhaps they are as clear.

No doubt the neglected discovery of Waller, again made and successfully propounded by Cohnheim, that white corpuscles pass through the walls of uninjured capillary vessels into the tissues, was one which upset previous notions, and might well create in some minds a doubt concerning the doctrine of Virchow's " Cellular Pathology." But looking at the subject with the advantage of the five years which have elapsed since Professor Stricker wrote his article, one cannot doubt that the real state of matters is simply this :—that amœboid connective-tissue-corpuscles and white blood-corpuscles are all one set of bodies, though the first are in the tissues, and the others floating free in the blood ; and they might well be termed common or undifferentiated corpuscles. While there need be little doubt that pus-corpuscles are derived from white blood-corpuscles, there also need be no more doubt that they likewise originate both fissiparously and endogenously in the tissues.

I have used the expression "common or undifferentiated corpuscles;" and on this subject it may be necessary to make some remark. The vital powers may be enumerated as irritability or sensibility, contractility, nutrition or elaboration of substance, and reproduction. All these properties are possessed by an amœba and by amœboid corpuscles. But every separate living organism does not possess them all; there are individuals without reproductive power. In like manner, all the vital units do not possess throughout life all vital properties ; but *in the process of differentiation one property becomes exalted, while another is lost.* Muscular fibres, nerve-corpuscles, the corpuscular elements of peripheral nerve-terminations, and secreting corpuscles illustrate this. All of them have lost the reproductive power ; muscular fibres have exalted contractility, nerve-terminations exalted sensibility, and secreting corpuscles a highly developed elaborating power. [In brain-corpuscles the proliferating power may reappear in pathological circumstances, but it does so at the expense

of the specially developed properties.   In muscular
fibre it appears to be lost altogether.]

Probably the greatest difficulty in conceiving
of the origin of differentiated textural elements
from common corpuscles is to settle the relation of
epithelial to other corpuscles, and on that subject
it is not easy to give an opinion.   In particular,
the phenomena of skin-grafting, including the
stimulus given to the growth of skin over a whole
ulcer by the presence of grafts of minute size,
might even suggest the possibility of a sexual
distinction between the corpuscles of the graft and
those among which it is planted.   At all events,
the microscopy of skin-grafting is worthy of study,
and the utility of the practice affords evidence that
all the less differentiated corpuscles are not capable
of producing, at least without assistance, all other
kinds of corpuscles.   [Evidence of this is especially
to be seen in the circumstance that the most
effective grafts are those which completely disappear before being succeeded by a transparent
pellicle which spreads ; and also in the well-ob-
served fact that improvement in an ulcer to which
grafts have been successfully applied is not confined to the spots where they have been placed.]

It must be kept in mind that the corpuscular
mass of the embryo becomes early divided into

layers, of which the outer and the inner may be said to be opposed to the middle one, in respect that those become epithelial, while this becomes the source of other tissues. Further, the important observation by His, of the abundant cell proliferation at the circumference of the area vasculosa, and of the intrusion of the elements so formed into the interior of the embryo, must be kept in mind as giving rise in early embryology to a primary division of the corpuscular masses into centrifugal and centripetal. That His is right in his view that the whole skeleton and connective tissues, as well as the blood and blood-vessels, are derived from the centripetal group I cannot believe, because it appears to me that Remak's view of the origin of the skeleton corresponds much better than that of His with what may be seen in transverse sections of embryos ; but it does seem very possible that the centripetal development, being the source of the blood and blood-vessels, furnishes not only nourishment to them but corpuscles which conjugate with all those of the centrifugal mass. I throw this out as a suggestion, and it will not, perhaps, be considered a very wild one, when this remark of Stricker's is remembered :—" Were any one to maintain that the migrating cells are conjugation organisms, no stronger objection could

I

be raised against him than against another who should maintain that the migrating cells are epithelia. Recklinghausen has advanced a theory respecting the conjugation of cells, which, however, on account of its brevity, scarcely allows us to judge of its value." The theory of Recklinghausen refers to conjugation between different elements of blood.

The part taken by Dr. Beale in the advance of the cell conception has been one of great importance, and is worthy of full consideration. The great merit of Dr. Beale appears to me to lie in pointing out that the cell-wall is in all instances an after-growth, and that the vital processes of the corpuscle are independent of it. In saying this, I am not forgetful of the work of Max Schultze in 1863, quoted by Stricker; but Dr. Beale has the priority, and lays down the doctrine of the non-vital character of the cell-wall in a very clear and emphatic manner, classifying, as was not done before, cell-walls with the inter-corpuscular substance. He simplified thereby the conception of cell multiplication; for if the cell-wall be, even when present, no part of the vital corpuscle, there is no radical distinction between fissiparous and endogenous reproduction. His distinct recognition also that the origin of inter-communicating processes of cells is not by outgrowth of processes, but by the separation of corpuscles which gradually part, is most important ; though there is grave reason to pause before denying with him that continuity of structure is ever the result of separate elements sending out processes which unite.

But, however much it may be desired to give Dr. Beale full credit for the advances which he has made, it is easy enough to understand why that credit should be sometimes withheld, when one considers how these advances have been mixed up with a theory of " germinal and formed matter " which has made but little way. It is quite impossible to support the doctrine that all " formed matter " was once " germinal," particularly if such things as the matrix of cartilage and the fibres of tendon are to be included under the term, as they are by Dr. Beale. Certain cell-walls, as those of at least some of the fat-cells, are really altered protoplasm ; but there is not the slightest reason to believe that the matrix of cartilage or the fibres of tendon are transformed portions of the vital corpuscles, or that they are undeserving of the name of " intercellular " substance. Rather would it have been well if Dr. Beale had looked on the cell-wall itself as intercellular.

His difficulty appears to be that " no well-

ascertained facts have yet been adduced in favour
of the view that any living structure whatever can
influence matter at a distance from it, so as to alter
its properties or composition, or in support of the
notion that cell-wall, cell-contents, or intercellular
substance possess any metabolic power whatever."
And the way he gets over this difficulty in the case
of hepatic cells is most ingenious, namely, by re-
presenting that the outer part of each corpuscle is
no longer vital, but converted into formed material
of a soft description, becoming changed into bili-
ary constituents, albuminoid and amyloid matters.
Other researches, however, come to our rescue here.
There is nothing more certain than the power
which Chrzonszczewsky showed hepatic corpuscles
to possess of taking up sulpho-indigotate of soda
from the blood and passing it on into the gall-ducts.
Here, then, is an indubitable instance of a living
structure influencing matter external to it. And
where, after all, is the unaccustomed marvel in this,
when it is recollected that all the attractions of dead
matter are exercised by molecules or masses, as the
case may be, on others external to themselves, and
that in the case of gravitation there is no limit to
the distance which may be between the masses
provided that they are sufficiently large ?

Looking at things from my point of view, I am

also obliged to think it a pity that Dr. Beale has not recognized the true place of muscle in his theory. He calls the contractile substance of muscle "formed matter," which indeed it is in the sense of being raised to a higher state of organization than the corpuscles out of which it is formed ; but the formed matter of Dr. Beale is, according to his definition, no longer vital ; and that is not the case with muscular fibre, which not only has the vital power of contractility, but the power of consuming other than its own substance in production of contraction. The real definition of a striped muscular fibre is, that it is a compound living corpuscle which has no reproductive power, but has a far more highly developed contractility than the amœboid corpuscles.

It may be stated that at the present day there is no difficulty in believing in the uninterrupted sequence of corpuscles by reproduction through all generations. The melting of the spermatozoa and germinal vesicle within the ovum may be regarded as a variety of conjugation, resulting in the formation of a corpuscle, which by its fissiparous division constitutes the wholly corpuscular germinal membrane, from which are furnished the parents of every living corpuscle in the adult body, including the ova or the spermatozoa, according to the sex.

But the mere tissue-life in individual corpuscles will not account for the phenomena of development without the addition of a larger life or a formative principle common to the whole individual, and it would be of incalculable advantage in the just conception of pathological phenomena, if the central and tissue lives were more generally distinguished than they are. [No one has yet reduced, in a satisfactory way, any of the properties above mentioned as belonging to corpuscles, namely, irritability, contractility, nutrition and reproduction, to the laws of unorganized matter ; and having regard to that circumstance, and to the complicated phenomena of development of higher organisms, exhibiting series of changes unlike anything in the organic world, it is legitimate to conclude that in living beings there is a superadded element acting on the textural units individually, and that such an element controls likewise the development of the organism. The neoplasms of the pathologist afford abundant example of corpuscular life breaking loose from the central control by means of which it is utilized in health for the construction and continuance of definite organs.]

Still proceeding on the principle of life within life, we may go further and assert that a larger life, or series of developmental changes from a simple origin to a definite goal, may be observed in the evolution of all animal forms in the history of the globe. Such a doctrine alone is capable of explaining all the facts of morphology, and giving to the speculations of Darwin the backbone which they require. [The reasons for it are stated in the foregoing essay on the Evolutions of Organization. It may be added here that such a doctrine is favourable rather than adverse to the supposition of a genetic relationship of widely separate species. For if not a textural unit has arisen save by genesis from pre-existing units, and not an organism save from pre-existing organisms, analogy, which is however an uncertain guide, suggests that not a species has arisen save from pre-existing species. If the pedigree of man from inorganic matter has during long ages passed through an ascending series of forms, so also has individual manhood passed through stages of development equally inferior to the adult condition. But we enter on another inquiry when we ask by what means the highest development of organization has been evolved both in the individual and in the world-history of life, even supposing that genetic relationship of all the parts be admitted as probable.]

VI.

TRUTH, PATHOLOGY, AND THE PUBLIC.

VI.

TRUTH, PATHOLOGY, AND THE PUBLIC.

Address delivered at the Ceremony of Graduation in the University of Glasgow, July 1880.

Gentlemen, the duty devolves upon me to address to you a few words of congratulation on the present auspicious epoch of your lives. Auspicious may it prove to all of you. To all it must be a satisfaction that you no longer have examinations to look forward to, and no longer feel obliged to read wearily for such ordeals, instead of studying for the sake of information. It is to be hoped, therefore, that your days of study are not now over, but only about to begin, and that the education which you have received at College will prove principally useful by teaching you the methods to be pursued in acquiring knowledge during the rest of your lives.

All education proceeds on one of two plans—the pin-cushion plan, which regards the mind as a dead receptacle, to be packed with facts and fictions, useful and otherwise; or the horticultural plan, which looks on it as a living plant, to be nursed to a healthy and more mature condition, ere the time for transplanting arrives, when it is to become useful or ornamental by the exercise of the healthy powers which have been educed. It is palpable that in all professional education there is a large mass of fact to be learned, but obvious also that the facts will be of little use without the art of handling them and turning them to account. More especially is this the case in medicine. The facts which are brought under the notice of medical students are, as you have experienced, these: the modes of preservation of health and treatment of disease; the laws of health and disease and the actions of remedies, which form the immediate basis of treatment; the structure of the body, without a knowledge of which its operations, healthy and otherwise, cannot be known, and without which you dare not use the knife; the chemical laws which govern both the body in its operations and the remedies which you propose to use; and lastly, the characters of animals and vegetables, that domain of life to which our life belongs. I have placed foremost that which is most prominent in the minds of most of you—the treatment of disease; and I do not doubt that you have all learned a great number of details of treatment which you will use to the advantage of patients yet unborn, as well as others, it is to be hoped. But remember that the changes which take place in treatment are perpetual, and that there is nothing reliable in any treatment which is not based on science.

The jeers of Le Sage and Molière, and many a sneer of later date, have been only too well founded; and if there be truth, as truth there is, in the vaunted progress of medicine and surgery in recent years, that progress is entirely owing to two closely connected causes, namely—first, the enormous advances that have been made in chemistry, natural history, anatomy, and physiology; and, secondly, that the practitioner, prepared by the study of these sciences, has applied their methods in his own special studies, has founded a science of pathology, and learned what accuracy of observation means in clinical research. It is then the methods learned in your scientific studies which those of you who aim at reaching the first rank in our profession will find the most useful part of your education, as you woo Nature for her hidden treasures all your lives. This lotion and that operation, O young surgeons; this plaister and that panacea, most promising physicians, will be superseded, it may be, in a very few years ; but if you have learned the scientific method, you will help, by patient and accurate observation and thought, progress which shall endure when much crude innovation known by the name shall have been abandoned and forgotten.

How hard it is to be accurate! Nay, rather how impossible! Accuracy is approached as if by a process of dividing the distance. Constant effort produces constant progress ; but a fraction of the distance continually remains. At least, so it is in matters of observation and construction. We begin them with a preconceived notion as to the degree of exactness required ; as we proceed we find we have to amend our notions; and when we have done this several times we find it exceedingly difficult or even impossible to recall the state of mind from which we started. Turn back your minds to the first few days of your studentship, and try to realise your first impressions of, for example, a verte-bral column. Probably you thought that nothing could be easier to understand, but wondered at the tiresomeness of detail in the descriptions given by authorities. You will probably also recollect that your teachers took a great deal of trouble to con-vince you of the importance of much which you

had difficulty in appreciating when pointed out to
you, simply because at that stage of your development you thought it far too minute and trifling for
a rational being to attend to.

Without going beyond the study of anatomy I
might point out many more striking illustrations of
the difficulty of arriving at accuracy, by asking you
to look back, not on your own history, but on
history which can be studied from records, the
progress of the science. One example will suffice.
Consult the plate showing the vascular system
originally published by Vesalius in 1542, and you
cannot fail to be astonished at its extraordinary
character. The inextricable confusion which it
exhibits in matters long regarded as fundamental
almost surpasses comprehension ; and yet Vesalius
was far ahead of his contemporaries. He was
conscientiously, and lovingly describing what he
believed that he had seen ; and the name that he
has left behind him, as well as the remarkable story
of his life, is guarantee for the ability which he
brought to his work. In 1690, appeared Bidloo's
work with its celebrated engravings ; and in the
plate of the arteries there given, not only is the
influence of the plate by Vesalius abundantly
evident, but in some respects the inaccuracies are
distinctly greater, though as a work of art it is

beautiful. It is more difficult to conceive the point of view of these old anatomists than to remember the changes which have gradually taken place in our own conceptions of accuracy; and yet in this instance we have but an example of what may be often seen both in natural history and in human affairs, that the development of the individual repeats after a fashion the development of the race. The art of accuracy had to be learned in anatomy and other sciences precisely as each of us has had to learn it, and as we shall continue to learn it as long as we admit the paramount excellence of truth.

Nor is that which obtains with regard to accuracy in scientific matters less applicable or important in reference to the statements which we allow ourselves to make in the ordinary conversation and business of life. All truthfulness is an art, and a difficult art to learn. In scientific matters the only difficulty in learning this art in most cases is to discipline the observation and teach it to act with rigour and free from prejudice and imagination. But in the affairs of life, and especially in our profession, there are more serious difficulties in the practice of truthfulness than these. Emotion comes in—emotion, which into scientific investigation ought not to enter though it too often does; and I do not know a

single emotion that has not got a warping power.[1] In
your practice you will make blunders, everybody
does ; and you will try to conceal them, everybody
does ; it will often be well for your patient that you
should. You will learn much which your patients
will think that they have a right to know, but which
it would be wrong to tell them ; and patients'
friends will ask distressing questions which it may
be your duty not to answer. They will ask you
questions, also, especially in the early days of your
practice, which you may find eminently inconvenient
to yourselves to answer, even when you know full
well that you are thoroughly able to do your duty.
In all these cases, and in many more which might
be mentioned, you will readily see how emotion of
one sort or other may interfere with accuracy of
statement ; and with regard to them all, I may
further say, happy is the man who can bring tact to
his rescue or even finesse in aid of truth.

[1] It is indeed true that all emotion has a warping power; yet I cannot republish this sentence without noting that I have fallen into a
specious error in saying that emotion ought not to enter into scientific
work. The emotions should all be strictly disciplined, especially those
founded on personal pique, which was the emotion most present to my
mind. Even pique has warmed men up to work, though not of the
best sort. As for the man who brings little enthusiasm into his work,
he will take little credit out of it, either in the way of discovery or of
influencing other minds. Yet enthusiasm continually warps the judgment, and we must curb our enthusiasm for our own views of truth
by loyal adhesion to the spirit of candour and caution by which we
can alone hope to throw aside the prejudices to which we are liable.

K

No doubt these qualities are too often thought to mark their possessor as false, and it is beyond question that they may be used in support of falsehood as well as in aid of truth.   Yet there is a form of blarney to be found in all grades of society perfectly legitimate when it is the offspring of geniality.   It is an art not unknown to the fashionable practitioner; perhaps you may find occasion to practise it.   Be tender at least with feelings that are tender and sore, and ever when inclined to prattle remember and beware.   It is the man who is looking out with clear and honest eyes, not thinking of himself, that has most time to study his neighbours and become a true tactician; while they who rush inconsiderately into every false position have most difficulty in emerging with clean hands; and the tongues which go spinning for ever like tops are they the burden of whose discourse is most frequently an hum.   It is the pace that kills.   Of such an one, had not Johnson objected to playing upon words, it might be said in a sense far different from what he intended, " Nullum quod tetigit non ornavit," for he adds legs and arms to every story he takes up.

Am I wrong then in considering honesty an art, and a very difficult one too?   "An honest man's the noblest work of God," says Pope in his "Essay on Man," much to the disgust, as you recollect, of Mr. Burchell, who thought the reputation of men was " to be prized, not from their exemption from fault, but the size of those virtues they were possessed of," a sentiment in which there is much truth ; but the more you think over it the more you will find that perfect honesty is a virtue of such size, a gem so rare, that it is not to be found anywhere, but is to be perpetually and earnestly sought after. Professor Wendell Holmes, American anatomist, and most pleasant of writers, tells us how comes " Timidity, and after her Good-Nature, and last of all Polite Behaviour, all insisting that truth must *roll*, or nobody can do anything with it ; and so the first with her coarse rasp, and the second with her broad file, and the third with her silken sleeve, do so round off and smooth and polish the snow-white cubes of truth, that when they have become a little dingy by use, it becomes hard to tell them from the rolling spheres of falsehood."

If we consider the three learned professions we shall find that in all of them there are incentives to truthfulness, and peculiar provocations to the reverse. The lawyer has often to practise special pleading, which may possibly, in some instances, weaken the moral antipathy to wrong ; but he has to study from the commencement the abstract principles of law, and the phases of its development, and learns as a judge to balance evidence independent of emotion.   Clergymen, in the individual exercise of their profession, have before them the highest ideal ; but bearing on them is the unhappy fact that in ecclesiastical organizations, as such, abstract truth or error has no *locus standi*, but every consideration yields to the question of accordance with the laws of the church.   As to medicine, I have already touched on the difficulty of being truthful in the practice of the profession ; but, on the other hand, in the search after general laws, authority is now-a-days armed with less persecuting power than formerly, and freedom is enjoyed in scientific inquiry.

Returning then to the subject of scientific accuracy : the physician is set face to face with nature ; and all correct treatment of disease must be founded on an accurate knowledge of pathological changes.   If your practice be not so founded, it will be quackery ; and the public and the state are to blame that the opportunities of arriving at a correct pathology are not what they ought to be.

There are two senses in which pathological anatomy is the foundation of correct practice.   In the first place, the profession, as a profession, could have no knowledge at all of any pathological change without studying morbid conditions after death. That is to you self-evident; but the general public have never caught hold of the idea properly, or they would take more care that the greatest possible facilities should be given for the prosecution of inquiries on which their own health ultimately depends. But let the public listen to this further statement, which you will all recognize as true, .though some of you will forget it as time rolls on— that the practice of each individual practitioner is trustworthy only when he makes use of frequent opportunities of examining after death to verify and supplement his judgments where they have been right, to test his guesses and his suppositions that he may see how far they have accorded with fact, and most of all to correct the numerous errors into which the wisest and the most experienced continually fall.

We talk of the progress of medicine; but what an enormous amount of distress might be alleviated if that which is known could be brought to bear in every instance in which it is applicable, or even in the majority of instances. To this end continued pathological observation must be conducted by every practitioner; and it is because the blame of the neglect of this lies much more with the public than with the profession that I venture to point out what happens when verification of diagnosis is neglected. Let us try to trace the history of a medical career in which this fault is committed. A graduate has been accustomed in his studies to have series of cases brought before his notice, of which the most prominently impressed on him have got well, and have seemed thereby to show that the evil had been accurately recognized and successfully combated. Others, likewise distinctly remembered, have got worse and worse, till they have ended fatally; and in those cases a proper investigation afterwards has taught numerous lessons which could not fail to impress the thoughtful observer. And if, besides all those, there have been, as there must be, numerous other cases which have not been rounded off to a dramatic conclusion of success or tragic close, but have lingered on in an unsatisfactory way, or disappeared from observation none the better; the exigencies of teaching, apart from the operation of any subtle and unconscious instinct of human nature in the mind of the teacher, tend often to throw such cases into the background, while their absence of sensational interest leads the student to leave them there to be forgotten. The graduate passes, as many of you are about to do, into general practice on his own account. Then how great is the change! The acute cases which

were so instructive in hospital no longer form the
staple of his experience, and the chronic cases no
more like meteors disappear from view. He has
these disadvantages to cope with—namely, that in
his successful cases, as will always happen, he is in
a certain number of instances deceived by the circumstance that the fortunate issue has not in reality
been aided by his treatment, or that the treatment
has accidentally suited a condition other than the
presumed lesion against which it was directed ; in
fatal cases he seldom has the opportunity afforded
him of proving the opinions on which he has acted,
and when the opportunity occurs he learns to consider it a trouble to utilize it; while, as for the vast
number of cases that seem to get neither greatly
better nor worse, they become inevitably, as time
goes on, a lighter burden on his conscience, are set
down as among the things that " no fellow can find
out," and kept as quiet as may be with a variety of
palliatives. What wonder that such a man falls
into a humdrum routine !

Suppose, in obscure cases, you exercise your
reasoning powers to the best of your ability ; you
form the theory which seems most probable ; but
you have no means of testing it ; other cases occur
similar in much to those which have been already
seen, and you apply the unproven theory with all

the strength of conviction, and so keep adding to what you call your experience—never verifying, and perhaps altogether wrong in your ideas as to what had really happened in the interesting cases on which you have built. Will it be remarkable if your notions become crude, if your own belief in pathology and in precise diagnosis become dulled, and your practice degenerate to a stupid superstition? This picture is by no means an altogether imaginary one. It is the record of what I have seen. The cure is to be found in the increase of facilities for comparing the theories formed in attendance on the living with the revelations that are offered by the dead; and that is a matter which rests with the general public.

The general public is a vague and not altogether satisfactory body to deal with. It is possible to have the greatest respect for your fellow-men individually, and have little for the general public: and justly so; because the opinions, passions, and prejudices that pass most current in the throng are by no means the secretly cherished and better judgments of the individuals who compose it. We do not air most blatantly on all occasions the wisdom which we privately think the best; I fear we should be prigs if we did; but so it happens that it is something a good deal worse than our best by which in our social relations we influence others and are influenced. We must also take into consideration that when we analyse the characters of individual men, we find that no man is wise in everything, and that most are wise in reference to things within a very limited compass. Yet, as it is the property of truth to keep ever sapping the domain of error, there is an always increasing number of truths and half-truths which gain the assent of the majority of men. From such considerations as these, it arises that public opinion is often the foolish offspring of passion and prejudice, especially with regard to mere passing events ; and yet, with regard to permanent matters, it keeps slowly and lumberingly, but very surely, moving on at a respectful distance behind the heels of wisdom. We are bound, therefore, in those professional matters with regard to which it would be better for the public to be liberated from prejudice, and be more fully informed, we are bound to keep hopefully pointing out the things which our training brings specially under our notice, and so endeavour by the one thing in which we are wise to make up for the many in which we are foolish.

As to the propriety of the public affording facilities to medical men to do their duty in the matter of making certain the cause of death in every case, it is to be frankly admitted with satisfaction that, in this, Scotland is far ahead of some other parts of the British Isles. I doubt, however, if there is anything like a proper conception of the fearful way in which the public suffers when the duty in question is neglected by medical men.

The public has to learn that an unskilful practitioner is simply a waster of human life, and that the public itself is but a bad judge of the waste of its own life which may take place. The popular prejudice against a custom thousands of years old in royal families is the cause of an enormous annual slaughter among all ranks, not the less real because it is impossible to compute.

It may be further mentioned, that it is at least supposed that the government statistics, made at considerable expense, of the causes of mortality, are of some use; yet it is an obvious fact, of which I have had abundant experience, that the returns which medical men have, under compulsion, filled up, stating causes of death, are in large part utterly worthless, not from any wilful dereliction of duty on the part of members of the profession, but because information is asked which they are not in a position to give. The schedules sent out by the registrar for certification of cause of death ought to demand whether or not a post-mortem examination has been made, and the sanitary department ought to keep the results obtained in cases where such examination has taken place, completely separate from the comparatively worthless returns of cases in which there has been none.

Moreover, when it is considered what pains are taken to secure accurate returns of causes of death in the army, and when it is remembered what a very large number of the practitioners engaged in civil. practice hold appointments in a great variety of public institutions, one would think that it only required a proper knowledge of the issues at stake to rouse the public to the consideration of its own safety, and to extend in some measure military methods to civil practice, so as to develop a more thorough knowledge of their profession among all ranks of medical men, and save the lives of rich and poor.

I desire to use this opportunity to say one word about pathological museums. They are the means of furnishing permanent and accessible records of remarkable phenomena, and are even more useful in affording comparison of different stages and varieties of disease one with another. There are no doubt certain matters connected with the general appearance of morbid textures which are better observed in the recent condition ; but there is much more which cannot be seen to advantage, nor indeed studied at all, except in carefully designed preparations permanently preserved.

Now, the great promoters of the art of displaying structure without desiccation were William and John Hunter, two Lanarkshire brothers. The museum of John Hunter, the property of the College of Surgeons, London, is continually and abundantly added to in a manner which would be highly satisfactory to its founder. The museum of William Hunter, a wonderful collection, embracing books, engravings, paintings, coins, and objects of natural history, was bequeathed to this University, and finds in Professor Young an enthusiastic curator; but its circumstances are most unfortunate as regards its anatomical and pathological department, the department which is most closely connected with the reputation of its founder. This interesting collection, made by the elder of the two brothers, who had the merit of teaching to his younger brother the art, is not, like the collection of John Hunter, a living centre which gathers to itself the best of present work; and this University is at the present moment almost powerless to make it so. But there can be no doubt that Glasgow ought to be the seat of one of the finest, most actively increasing, and most useful pathological collections

in the world; and when I think of the public spirit
of this great city, and the enlightened character of
those who have it in their power to bring this
about, I cannot doubt that the time will rapidly
come when a great and successful effort will be
made by the supply of funds, and by proper combination, to give to a science so important for the
direct interests of the public as pathology is, the
support in Glasgow which it ought to enjoy. We
require much extension of accommodation in our
medical school, so rapidly do the wants of medical
teaching develop; and a special representation of
pathology on our staff is greatly to be desired, as
well as an united medical museum, to which not
only students may have liberal admission for pur-
poses of study, but medical men may resort from
all parts, both for their own immediate benefit and
also to help them in work by which they may
enlighten others.

In the path of a progress of which we can have
no fear as to whither it will lead we must not halt;
and it is with pleasure, therefore, that I take this
opportunity of attracting attention to an important
advantage added this spring to our medical school
by the appointment of the able medical superin-
tendent of the Royal Asylum of Gartnavel as
University Lecturer on Insanity. I trust that when

another year comes round our senior students will mark their appreciation of the important privilege thus offered them.

Graduates, on you we depend for further advance. It is for you, by earnestness and honourable work, to reflect new credit on your Alma Mater. I ask of you, wherever you go, to preserve for her a kindly regard. To you who go to foreign countries I would say, do not forget her museums; and all of you be assured that among your old teachers you will ever find a sincere desire to help you in the pursuit of knowledge.

GLASGOW :
Printed at the University Press
BY ROBERT MACLEHOSE, 153 WEST NILE STREET.

WORKS PUBLISHED BY

MR. MACLEHOSE,

Publisher to the University of Glasgow.

SAINT VINCENT STREET, GLASGOW.

Published by
JAMES MACLEHOSE, GLASGOW,
𝔓ublisher to the 𝔘niversity.

LONDON : MACMILLAN AND CO.

London,	.	.	.	*Hamilton, Adams & Co.*
Cambridge,	.	.	.	*Macmillan & Co.*
Edinburgh,	.	.	*Douglas & Foulis.*
Dublin,	.	*W. H. Smith & Son.*

MDCCCLXXXI.

MR. MACLEHOSE'S
Catalogue of Books.

ARGYLL, Duke of—WHAT THE TURKS ARE; AND HOW WE HAVE BEEN HELPING THEM. Speech delivered in the City Hall, Glasgow, September 19, 1876. By His Grace the DUKE OF ARGYLL, K.T. With a Preface. Second Edition. 8vo. 1s.

ARGYLL, Duke of—SPEECH ON THE CONDUCT OF THE FOREIGN OFFICE DURING THE INSURRECTION IN CRETE IN 1867. 8vo. 6d.

ANDERSON—CLINICAL LECTURES ON THE CURABILITY OF ATTACKS OF TUBERCULAR PERITONITIS AND ACUTE PHTHISIS (GALLOPING CONSUMPTION). By T. M'CALL ANDERSON, M.D., Professor of Clinical Medicine in the University of Glasgow. Crown 8vo, Illustrated with Wood Engravings. 2s. 6d.

ANDREWS—THE PSYCHOLOGY OF SCEPTICISM AND PHENOMENALISM. By JAMES ANDREWS. Crown 8vo. 2s. 6d.

BANNATYNE—GUIDE TO THE REGIMENTAL EXAMINATION FOR PROMOTION OF OFFICERS IN THE INFANTRY. Part I. —Containing what is required by Her Majesty's Regulations to qualify for promotion to the Rank of LIEUTENANT. By LIEUTENANT COLONEL BANNATYNE. Sixteenth Edition, carefully revised. 1880. Small 8vo. 7s.

BANNATYNE—GUIDE TO THE REGIMENTAL EXAMINATION FOR PROMOTION OF OFFICERS IN THE INFANTRY. Part II. Containing what is required by Her Majesty's Regulations, in addition to the subjects comprised in Part I., to qualify for Promotion to the Rank of CAPTAIN. Thirteenth Edition, carefully revised. 1880. Small 8vo. 7s.
[In preparation.

BANNATYNE — INSTRUCTIONS FOR THE PAYMENT OF TROOPS AND COMPANIES IN THE CAVALRY AND INFANTRY, founded on the Queen's and War Office Regulations. Small 8vo. New and greatly enlarged Edition. 6s.

BANNATYNE—BRIGADE DRILL, in the form of Question and Answer, founded on the Field Exercise and Evolutions of Infantry. Small 8vo. 1s.

BATHGATE—COLONIAL EXPERIENCES; OR SKETCHES OF PEOPLE AND PLACES IN THE PROVINCE OF OTAGO, NEW ZEALAND. By A. BATHGATE, Dunedin. Crown 8vo. 7s. 6d.

"Anybody who may be desirous of getting an idea of the real state of matters in New Zealand, cannot do better than read Mr. Bathgate's pleasant, chatty, unpretending colonial experiences."—*Leeds Mercury*.

BLACKBURN—THE PIPITS. By the Author of "Caw, Caw," with Sixteen page Illustrations by J. B. (MRS. HUGH BLACKBURN). 4to. 3s.

"This is a charming fable in verse, illustrated by the well-known J. B., whose power in delineating animals, especially birds, is scarcely inferior to Landseer or Rosa Bonheur."—*Courant*.

BORLAND HALL, by the Author of "OLRIG GRANGE." *See* SMITH.

BROWN—THE LIFE OF A SCOTTISH PROBATIONER. Being the Memoir of THOMAS DAVIDSON, with his POEMS and LETTERS. By the REV. JAMES BROWN, D.D., Minister of St. James' Church, Paisley. Second Edition. With Portrait. Crown 8vo. 7s. 6d.

" A charming little biography. His was one of those rare natures which fascinates all who come in contact with it."—*Spectator*.

" It is an unspeakable pleasure to a reviewer weary of wading through piles of commonplace to come unexpectedly on a prize such as this."—*Nonconformist*.

" A very fresh and interesting little book."—*Saturday Review*.

BUCHANAN—CAMP LIFE IN THE CRIMEA AS SEEN BY A CIVILIAN. A Personal Narrative by GEORGE BUCHANAN, M.A., M.D., Professor of Clinical Surgery in the University of Glasgow. Crown 8vo. 7s. 6d.

BUCHANAN—INAUGURAL ADDRESS. Delivered in the University of Glasgow. By PROFESSOR BUCHANAN. 8vo. 1s.

CAIRD, Principal—AN INTRODUCTION TO THE PHILOSOPHY OF RELIGION. By the VERY REVEREND JOHN CAIRD, D.D., Principal and Vice-Chancellor of the University of Glasgow, and one of Her Majesty's Chaplains for Scotland. Demy 8vo. 10s. 6d.

" It is the business of the reviewer to give some notion of the book which he reviews, either by a condensation of its contents or by collecting the cream in the shape of short selected passages ; but this cannot be done with a book like the one before us, of which the argument does not admit of condensation, and which is all cream.........The most valuable book of its kind that has appeared."—Mr. T. H. Green in *The Academy*.

" It is remarkable also for its marvellous power of exposition and graceful subtlety of thought. Hegel's solution of the problem of religion is at length adequately represented in English literature. Hegelianism has never appeared so attractive as it appears in the clear and fluent pages of Principal Caird."—*Spectator*.

" This is in many respects a remarkable book, and perhaps the most important contribution to the subject with which it deals that has been made in recent years."—*Mind*, October 1880.

"To many a student the reading of this book will mark an intellectual and spiritual epoch."—*The Nation* (New York).

CAIRD, Principal—UNIVERSITY SERMONS AND LECTURES. 8vo. 1s. each.

1. WHAT IS RELIGION? A Sermon preached before the University, at the Opening of the University Chapel. 1871.
2. CHRISTIAN MANLINESS. A Sermon preached before the University, at the Opening of the Winter Session, 1871.
3. IN MEMORIAM. A Sermon preached before the University on occasion of the Death of the Very REV. THOMAS BARCLAY, D.D., Principal of the University.
4. THE UNIVERSAL RELIGION. A Lecture delivered in Westminster Abbey, on the day of Intercession for Missions.
5. THE UNITY OF THE SCIENCES. A Lecture delivered at the Opening of the Winter Session. 1874.
6. THE PROGRESSIVENESS OF THE SCIENCES. A Lecture delivered at the Opening of the Winter Session. 1875.

CAIRD, Professor E.—A CRITICAL ACCOUNT OF THE PHILOSOPHY OF KANT : with an Historical Introduction. By EDWARD CAIRD, M.A., LL.D., Professor of Moral Philosophy in the University of Glasgow, and late Fellow and Tutor of Merton College, Oxford. 8vo. 18s.

"This book contains the most exhaustive and most valuable exposition of Kant's metaphysical system which has appeared in this country. The critical analysis is incisive and searching, and the exposition plain and unambiguous. The running commentary is the work of a man who both knows what he has to say, and knows how to say it forcibly and well. The style is attractive as well as clear. Without being ornate or rhetorical, it has about it a kind of quiet eloquence which comes of conscious strength and of genuine conviction."—*The Times.*

"Mr. Caird's statement of the Kantian doctrine is singularly felicitous. The simplification is at once full, accurate, and unbiased."—MR. T. H. GREEN in the *Academy.*

"No account of Kant's Philosophy has ever appeared in England so full, so intelligible, and so interesting to read as this work by Professor Caird. It is *the* English Book on Kant."—*Contemporary Review.*

CAMERON—LIGHT, SHADE, AND TOIL : POEMS. By WILLIAM C. CAMERON, Shoemaker. With Introductory Note by the REV. WALTER C. SMITH, D.D. Extra fcap. 8vo. 6s.

CLELAND—EVOLUTION, EXPRESSION, AND SENSATION, by
JOHN CLELAND, M.D., F.R.S., Professor of Anatomy in
the University of Glasgow. Crown 8vo. [*In the Press.*

DAVIDSON—LIFE OF—*See* BROWN'S LIFE OF DAVIDSON.

DEAS—HISTORY OF THE CLYDE TO THE PRESENT TIME.
With Maps and Diagrams. New Edition Enlarged. By
JAMES DEAS, M. Inst. C.E., Engineer of the Clyde Naviga-
tion. 8vo. 10s. 6d.

" Mr. Deas tells his story in a very clear and concise way."—*Saturday
Review.*

DERBY, Earl of—INAUGURAL ADDRESS, delivered to the
University of Glasgow, on his Installation as Lord Rector
of the University. By the Right Honourable The EARL
OF DERBY. 8vo. 1s.

DICKSON—PLEADING IN THE COURTS OF LAW OF SCOT-
LAND. An Address delivered before the Glasgow Juridical
Society. By W. GILLESPIE DICKSON, LL.D., Advocate,
late Sheriff of Lanarkshire. 8vo. 1s.

DICKSON—INTRODUCTORY LECTURE. Delivered at the
Opening of the Divinity Hall, in the University of Glasgow,
Session 1873-74. By WILLIAM PURDIE DICKSON, D.D.,
Professor of Divinity. 8vo. 1s.

DODS—ON PREACHING, an Address delivered to the Students
of the Free Church College, Glasgow, by Marcus Dods,
D.D. Second Edition, 8vo. 6d.

DOTTY AND OTHER POEMS by J. L. Extra Fcap. 8vo. 3s. 6d.

EGGS 4D. A DOZEN, AND CHICKENS 4D. A POUND
ALL THE YEAR ROUND. Containing full and com-
plete information for successful and profitable keeping of
Poultry. Small 8vo. Cheap Edition in Paper Covers.
Seventeenth Thousand. 1s.

"No book has appeared for many years that discusses the subject so
explicitly and exhaustively."—*Daily Review.*

EWING—MEMOIR OF JAMES EWING, ESQ., of Strathleven,
Formerly Lord Provost of Glasgow, and M.P. for that City.
By the late REV. MACINTOSH MACKAY, LL.D. Fcap. 4to.
With Portrait. 21s.

GAIRDNER—FLUCTUATIONS IN TRADE. A Lecture delivered
to the Institute of Accountants. By CHARLES GAIRDNER,
Manager of the Union Bank of Scotland. Third Edition.
Demy 8vo. 1s.

GAIRDNER, Professor—TWO LECTURES ON BOOKS AND
PRACTICAL TEACHING, AND ON CLINICAL INSTRUCTION.
Being Introductory Addresses delivered in the University
of Glasgow, and in the Western Infirmary. By W. T.
GAIRDNER, M.D., Professor of Practice of Medicine in the
University of Glasgow, Physician in Ordinary to Her
Majesty the Queen in Scotland. 8vo. 1s.

GEMMEL—THE TIBERIAD ; or, The Art of Hebrew Accen-
tuation. A Didactive Poem, in Three Books. By JOHN
GEMMEL, M.A., Senior Minister of Free Church at Fairlie.
Extra fcap. 8vo. 3s.

GLASGOW—THE OLD COUNTRY HOUSES OF THE OLD
GLASGOW GENTRY. Illustrated by permanent Photographs.
Royal 4to. Half Red Morocco. Gilt Top. Second Edition.
Very scarce. £10, 10s.

This is a history of one hundred of the old houses in and around
Glasgow, and of the families who owned and lived in them. To the
local antiquary it is especially interesting as a memorial of the old burgher
aristocracy, of their character and habits, and of the city in which they
lived ; while to the descendants of the "old gentry" it is interesting as
containing the history of their forefathers and the rise of their families.

GLASGOW—MEMORABILIA OF THE CITY OF GLASGOW Selected from the Minute Books of the Burgh, 1568 to 1750. Fcap. 4to. Half Morocco, Gilt Top. *Very scarce.* 63s.

GLASGOW UNIVERSITY ALBUM FOR THE NEW COLLEGE. With a Photograph of the New College. Edited by Students of the University. Crown 8vo. 7s. 6d.

GLASGOW UNIVERSITY LOCAL EXAMINATIONS. Scheme of Examinations for 1881, and Report for 1880. Crown 8vo. 6d.

GLASGOW UNIVERSITY CALENDAR FOR THE YEAR 1880-81. Published annually. Contains official information as to the University, the Classes, Graduation in all the Faculties, Examination Papers, University Fees, Bursaries, Scholarships and Fellowships, and the list of Members of the General Council of the University. Crown 8vo. 2s.

GOVETT—CHRIST'S RESURRECTION AND OURS ; OR I. CORINTHIANS XV. EXPOUNDED. By the REV. R. GOVETT, Norwich. Crown 8vo. 3s. 6d.

GRANT—CHRISTIAN BAPTISM EXPLAINED. By the late REV. WILLIAM GRANT, Ayr. 16mo. 1s. 6d.

GRANT—THE LORD'S SUPPER EXPLAINED. By the late REV. WILLIAM GRANT, Ayr. Seventh Edition. 16mo. 4d.

GRAY, David—THE POETICAL WORKS OF DAVID GRAY. Edited by HENRY GLASSFORD BELL, late Sheriff of Lanarkshire. New and enlarged Edition, extra fcap. 8vo. 6s.

" Gems of poetry, exquisitely set."—*Glasgow News.*

JACK—INAUGURAL ADDRESS. Delivered in the University of Glasgow, November 1879. By WILLIAM JACK, M.A., LL.D., Professor of Mathematics in the University of Glasgow. 8vo. 1s.

JEBB—The Anabasis of Xenophon.—Book III., with the Modern Greek Version of Constantine Bardalachos (for the use of the Greek Classes in the University of Glasgow), and with an Introduction by R. C. Jebb, M.A., Professor of Greek in the University of Glasgow. Crown 8vo. 2s.

LEITCH—Practical Educationists and their Systems of Teaching. By James Leitch, late Principal of the Church of Scotland Normal School, Glasgow. Crown 8vo. 6s.

"This capital book presents us, in a compact and well-digested form, with all that is of most value in the really practical methods of the greatest educationalists."—*School Board Chronicle.*

LEISHMAN—A System of Midwifery, Including the Diseases of Pregnancy and the Puerperal State. By William Leishman, M.D., Regius Professor of Midwifery in the University of Glasgow. In one Thick Volume. Demy 8vo. Nearly 900 pp., with 210 Wood Engravings. Third Edition, revised. 21s.

"Unquestionably the best modern book on the subject in our language."—*British and Foreign Medico-Chirurgical Review.*
"It is the best English work on the subject."—*Lancet.*
"We should counsel the student by all means to procure Dr. Leishman's work."—*London Medical Record.*
"In many respects not only the best treatise on the subject that we have seen, but one of the best treatises on any medical subject that has been published of late years."—*Practitioner.*

MACELLAR—Memorials of a Ministry on the Clyde. Being Sermons preached in Gourock Free Church. By the late Rev. Robert Macellar. With a Biographical Notice by the Rev. A. B. Bruce, D.D., Professor of Theology, Free Church, Glasgow. With Portrait. Crown 8vo. 7s.

MACEWEN—Sermons. By the late Alexander MacEwen, M.A., D.D., Minister of Claremont Church, Glasgow. Edited by his Son. With a Memoir. Crown 8vo. 6s.

MACGEORGE—Papers on the Principles and Real Position of the Free Church. By Andrew Macgeorge. 8vo. 6s.

MACLEOD, NORMAN, D.D.—Sermons on the occasion of his Death, preached in the Barony Parish Church, and the Barony Chapel, by the Rev. Dr. Watson, Dundee ; Rev. Dr. Taylor, Crathie ; Rev. Mr. Grant, St. Mary's, Partick ; and Rev. Mr. Morrison, Dunblane. 8vo. 1s.

M'KINLAY, J. Murray—Poems. Extra fcap. 8vo. 3s. 6d.

M'KENDRICK—Outlines of Physiology, in Its Relations to Man. By J. Gray M'Kendrick, M.D., F.R.S.E., Professor of Physiology in the University of Glasgow. Crown 8vo. 750 pages, and 250 Engravings. 12s. 6d.

"We have much pleasure in confidently recommending this work to students of medicine and others, as being the one of all others of recent date, best suited for their requirements."—*British Medical Journal.*
"An admirable book on physiology, well adapted to the wants of the student, and of practitioners in medicine. Such books as this ought to be read, not alone by medical and biological students, but by all men of any pretensions to general culture."—*British Quarterly Review.*
"The style is clear, and the illustrations numerous."—*Practitioner.*

M'KENDRICK—Lectures on the Graphic Method of Physiological Research. Illustrated with numerous Engravings. Crown 8vo. [*In preparation.*

MACMILLAN—Our Lord's Three Raisings from the Dead. By the Rev. Hugh Macmillan, LL.D., F.R.S.E., Author of "Bible Teachings in Nature." Crown 8vo. 6s.

"He has written a book on one of the most trying themes, which is at once edifying and instructive, full of devotional fervour as of fine thought." *Nonconformist.*
"A spirit of earnest piety pervades the book ; its language is simple and unaffected, and it abounds in apt and felicitous illustrations."—*Scotsman.*

MARTYNE—Poems. By Herbert Martyne. Extra fcap. 8vo. 6s.

MAXWELL—Address to the Senatus and General Council of the University of Glasgow. Delivered at his Installation as Chancellor of the University, on the 27th April, 1876. By the late Sir William Stirling Maxwell, K.T., Bart., M.P. 8vo. 1s.

MORTON—The Treatment of Spina Bifida by a New Method. By James Morton, M.D., Professor of Materia Medica, Anderson's College, and Surgeon and Clinical Lecturer of Surgery in the Glasgow Royal Infirmary. Crown 8vo. With Six Illustrations. 5s.

MUIR—The Expression of a Quadratic Surd as a Continued Fraction. By Thomas Muir, M.A., F.R.S.E., Examiner in Mathematics in the University of Glasgow. 8vo. 1s. 6d.

MURRAY—Old Cardross, a Lecture. By David Murray, M.A., F.S.A. Scot. Crown 8vo, 1s. 6d. Large paper copies, on Dutch Paper, Parchment Wrapper, 6s.

NEWTON—Sir Isaac Newton's Principia. Edited by Sir William Thomson, D.C.L., LL.D., F.R.S., Professor of Natural Philosophy in the University of Glasgow, Fellow of St. Peter's College, Cambridge; and Hugh Blackburn, M.A., late Fellow of Trinity College, Cambridge. Crown 4to. 31s. 6d.

" So far as we have compared it with other copies, this edition seems to be better than any of its predecessors. The printing and paper are excellent, and the cuts especially are greatly improved."—*Nature.*

NICHOL—Tables of European Literature and History, from A.D. 200 to 1876. By John Nichol, M.A., Balliol, Oxon., LL.D., Regius Professor of English Language and Literature, in the University of Glasgow. 4to, Cloth. 6s. 6d.

" The tables are clear, and form an admirable companion to the student of history, or indeed to any one who desires to revise his recollection of facts."—*Times.*

NICHOL—TABLES OF ANCIENT LITERATURE AND HISTORY, FROM B.C. 1500 TO A.D. 200. By PROFESSOR NICHOL. 4to, Cloth. 4s. 6d.

" They constitute a most successful attempt to give interest to the chronology of literature, by setting before the eye the relation between the literature and the practical life of mankind."—*Observer.*

NICHOL—A NEW VOLUME OF POEMS. [*In preparation.*

OLRIG GRANGE. *See* SMITH.

PORTER—CHRISTIAN PROPHECY : OR POPULAR EXPOSITORY LECTURES ON THE REVELATION TO THE APOSTLE JOHN. By S. T. PORTER. Post 8vo. 7s. 6d.

PORTER—THE LAST SERMONS in a 41 Years' Ministry, and in the 24th Year of Pastorate in the Independent Church, in West Bath Street, Glasgow. By S. T. PORTER. Crown 8vo. 1s. 6d.

PULSFORD—SERMONS PREACHED IN TRINITY CHURCH, GLASGOW. By WILLIAM PULSFORD, D.D. Crown 8vo. Cloth, Red Edges. Cheap Edition. 4s. 6d.

" The sermons have much of the brilliancy of thought and style by which Robertson fascinated his Brighton hearers."—*Daily Review.*
" The preacher we are made to feel, speaks to us out of the fulness of his own spiritual and intellectual life. He is a preacher, because he has been first a thinker."—*Spectator.*

RANKINE—SONGS AND FABLES. By WILLIAM J. MACQUORN RANKINE, late Professor of Engineering in the University of Glasgow. With Portrait, and with Ten Illustrations by J. B. (MRS. HUGH BLACKBURN). Second Edition. Extra fcap. 8vo. 6s.

" These songs are exceedingly bright, strong, and clever : quite the best we have seen for long. They are, in our judgment, far superior to those of Mr. Outram and Lord Neaves, and these are no contemptible singers. An admirable photograph is prefixed to the volume."—*Aberdeen Journal.*

SMITH, Walter C.—RABAN ; OR, LIFE SPLINTERS : a Poem.
By the Author of "Olrig Grange." Extra Fcap. 8vo.
7s. 6d.   [*This day.*

SMITH, Walter C.—OLRIG GRANGE : a Poem in Six Books.
Edited by HERMANN KUNST, Philol. Professor.   Third
Edition.   Extra fcap. 8vo.   6s. 6d.

"This remarkable poem will at once give its anonymous author a high
place among contemporary English poets.—*Examiner.*

"The most sickening phase of our civilization has scarcely been exposed
with a surer and quieter point, even by Thackeray himself, than in this
advice of a fashionable and religious mother to her daughter."—*Pall Mall
Gazette.*

"The pious self-pity of the worldly mother, and the despair of the
worldly daughter are really brilliantly put. The story is worked out with
quite uncommon power."—*Academy.*

SMITH, Walter C.—BORLAND HALL : a Poem in Six Books.
By the Author of "Olrig Grange." Second Edition.
Extra fcap. 8vo.   7s.

"Songs of exquisite beauty stud the poem like gems in some massy work
of beaten gold.   Original and vigorous thought, rare dramatic instinct, and
profound knowledge of human nature are embodied in poetry of a very
high class.   The poem is not only notable in itself, but full of splendid
promise."—*Scotsman.*

"Lyell's mother, stern and unrepentant, even in death, is a terrible portrait, in which we recognize the genius of the author of 'Olrig Grange.'
. . .   A style singularly brilliant and passionately fervent, a verse melodi-
ous and various in measure, a command of language, unusually extensive
and apt, and an exquisite sensibility to all natural loveliness." —*English
Independent.*

SMITH, Walter C.—HILDA ; AMONG THE BROKEN GODS :
a Poem.   By the Author of "Olrig Grange."   Second
Edition.   Extra fcap. 8vo.   7s. 6d.

"That it is characterized by vigorous thinking, delicate fancy, and happy
terms of expression, is admitted on all hands."— *Times.*

"A poem of remarkable power.   It contains much fine thought, and
shows throughout the deepest penetration into present-day tendencies in
belief or no-belief."—*British Quarterly Review.*

SMITH, Walter C.—BISHOP'S WALK ; and Other Poems.
Extra fcap 8vo.   2s. 6d.

STANLEY, Dean—THE BURNING BUSH. A Sermon preached before the Glasgow Society of the Sons of Ministers of the Church of Scotland. By the Very Reverend ARTHUR PENRHYN STANLEY, D.D., Dean of Westminster. 8vo. 1s.

STEWART—THE PLAN OF ST. LUKE'S GOSPEL ; A Critical Examination. By the REV. WILLIAM STEWART, M.A., D.D., Regius Professor of Biblical Criticism in Glasgow University. 8vo. 3s. 6d.

STORY—CREED AND CONDUCT : Sermons preached in Rosneath Church. By the REV. ROBERT HERBERT STORY, D.D., Minister of the Parish. Crown 8vo. 7s. 6d.

'' Characterized throughout by profound earnestness and spirituality, and written in a style at once graceful, clear, and nervous. Dr. Story has made a well-timed attempt to widen the theology, and at the same time to deepen and intensify the religious feeling of his countrymen."--*Scotsman.*

STORY—ON FAST DAYS ; With reference to more Frequent Communion, and to Good Friday. By the REV. ROBERT HERBERT STORY, D.D., Minister of Rosneath. 8vo. 1s.

"A thoughtful and earnest discussion of a most important question."— *Edinburgh Courant.* " A very able pamphlet."—*Glasgow Herald.*

VEITCH—THE HISTORY AND POETRY OF THE SCOTTISH BORDER, THEIR MAIN FEATURES AND RELATIONS. By JOHN VEITCH, LL.D., Professor of Logic and Rhetoric in the University of Glasgow. Crown 8vo. 10s. 6d.

'' This is a genuine book. We can heartily recommend it to three classes of readers—to all who have felt the power of Scott's 'Border Minstrelsy' (and who with a heart has not?), to all who care to visit and really to know that delightsome land, for no other book except the ' Border Minstrelsy' itself will so open their eyes to see it ; to all dwellers in the Borderland who wish to know, as they ought to know, what constitutes the grace and glory of their Borderland."—*Contemporary Review.*
'' We feel as if we were hearing the stories, or listening to the snatches of song among the breezes of the mountains or the moorland, under the sun-broken mists of the wild glens, or the wooded banks of the Yarrow or the Tweed."—*Times.*
" The fullest, most thorough, and most deeply critical work on Border history and poetry that we have."—*British Quarterly Review.*

VEITCH—Lucretius and the Atomic Theory. By Professor Veitch. Crown 8vo. 3s. 6d.

"We have read this little volume with no ordinary delight. We warmly recommend it."—*Nonconformist.*

VEITCH—Hillside Rhymes :

"Among the rocks he went,
And still looked up to sun and cloud
And listened to the wind."

Extra fcap. 8vo. 5s.

"Let any one who cares for fine reflective poetry read for himself and judge. Next to an autumn day among the hills themselves, commend us to poems like these, in which so much of the finer breath and spirit of those pathetic hills is distilled into melody."—*Scotsman.*

VEITCH—The Tweed, and Other Poems. Extra fcap. 8vo. 6s. 6d.

"Every page bears witness to a highly cultivated mind : every page is also marked by originality and a deep love for nature."—*Westminster Review.*

VILLAGE LIFE—A Poem.

"He seems to be a stranger ; but his present is
A withered branch, that's only green at top."

Extra fcap 8vo. 6s. 6d.

"These are simply the ripest notes that have appeared in Scotland for a time too long to calculate."—*Examiner.*

"A remarkable volume of poetry, which will be read by all who have any keen interest in the progress of English literature."—*Standard.*

WADDELL—Ossian and the Clyde, Fingal in Ireland, Oscar in Iceland ; or, Ossian Historical and Authentic. By P. Hately Waddell, LL.D. 4to. 12s. 6d.

WATSON—Kant and his English Critics, a comparison of Critical and Empirical Philosophy. By John Watson, M.A., LL.D., Professor of Moral Philosophy in Queen's University, Kingston, Canada. 8vo.   [*In the Press.*